T0283600

Power and Energy Engineering

Power and Energy Engineering

Edited by
Jackson Williams

Power and Energy Engineering
Edited by Jackson Williams
ISBN: 978-1-63549-230-9 (Hardback)

Published by Larsen and Keller Education,
5 Penn Plaza,
19th Floor,
New York, NY 10001, USA

Cataloging-in-Publication Data

Power and energy engineering / edited by Jackson Williams.
p. cm.
Includes bibliographical references and index.
ISBN 978-1-63549-230-9
1. Electrical engineering. 2. Power (Mechanics). 3. Electric power systems. 4. Power electronics.
5. Power resources. I. Williams, Jackson.
TK145 .P69 2017
621.3--dc23

The publisher's policy is to use permanent paper from mills that operate a sustainable forestry policy. Furthermore, the publisher ensures that the text paper and cover boards used have met acceptable environmental accreditation standards.

Printed and bound in the United States of America.

For more information regarding Larsen and Keller Education and its products, please visit the publisher's website www.larsen-keller.com

Table of Contents

Preface VII

Chapter 1 **Introduction to Power and Energy Engineering** 1
- i. Power Engineering 1
- ii. Energy Engineering 9
- iii. Electrical Engineering 11

Chapter 2 **Power Electronics: Features and Applications** 29
- i. Power Electronics 29
- ii. DC-to-DC Converter 47
- iii. Rectifier 54
- iv. Power Inverter 74

Chapter 3 **Methods of Generating Electricity** 90
- i. Static Electricity 90
- ii. Wind Turbine 98
- iii. Electromagnetic Induction 113
- iv. Photovoltaics 121

Chapter 4 **Electric Power System: An Overview** 143
- i. Electric Power System 143
- ii. Electric Power Distribution 153
- iii. Electric Power Transmission 161
- iv. Power-system Protection 183
- v. Transformer 188
- vi. Fault Current Limiter 211

Chapter 5 **Analysis of Power Engineering** 215
- i. Power-flow Study 215
- ii. One-line Diagram 219

Chapter 6 **Essential Aspects of Power and Energy Engineering** 221
- i. Electricity Generation 221
- ii. Efficient Energy Use 229
- iii. Alternative Energy 245

Permissions

Index

Preface

Energy engineering is a new and rapidly growing field of engineering. It is concerned with facility management, alternative energy technologies, energy services, environmental compliance, energy efficiency, etc. Power engineering is that branch of energy engineering, which deals with the transmission, utilization, generation and distribution of electric power and devices like transformers, generators, motors, etc. Themes included in this book enumerate the various ways in which this field has impacted our lives in an everyday level. This book is designed to provide students with the topics that are of utmost significance. It picks up individual branches and explains their need and contribution in the context of the growth of this subject. Coherent flow of topics, student-friendly language and extensive use of examples make this textbook an invaluable source of knowledge.

Given below is the chapter wise description of the book:

Chapter 1- Energy engineering deals with energy efficiency, environmental compliance, energy services and related fields. As a discipline it combines physics, math, chemistry and environmental engineering and is focused on increasing efficiency and developing alternate energy technologies. This introduction explores the topic of energy engineering and its subdivision, power engineering. This chapter is an overview of the subject matter incorporating all the major aspects of power and energy engineering.

Chapter 2- Power electronics is concerned with the conversion of power and the devices using it may be employed as switches or amplifiers, which help in power modulation and regulation. The chapter studies devices like DC- to-DC converter, rectifier and power inverter. The attributes and features of each have been discussed in detail to enable an insightful understanding of the complex subject-matter.

Chapter 3- The generation of electricity from sources like wind using wind turbines, sunlight using photovoltaic devices and magnets using electromagnetic induction has been explored in this content. The principle and methodology of these methods of generating electricity have been discussed in-depth. There is a section devoted to static electricity, its causes and applications as well.

Chapter 4- An electric power system, sometimes referred to as the grid consists of a network of electronic units that are used to supply, transfer and use electricity. In this chapter the reader is introduced to the key concepts of electric power system like electric power distribution, electric power transmission, power-system protection, transformer and fault current limiter. By focusing on these topics the content aims at increasing the reader's awareness about the complex topics of electric power system.

Chapter 5- Power engineering uses numerical analysis in the form of power-flow study to assess various parameters of AC current like voltage, real power, voltage angles and reactive power. It uses notations like one-line diagram to achieve this. The content of the chapter is designed to further the reader's comprehension of power systems. Power engineering is best understood in confluence with the major topics listed in the following chapter.

Chapter 6- Power and energy engineering concentrates on the branches of electricity generation, efficient energy use and alternative energy sources which form its core aspects. This chapter explores these aspects by elaborating on the history, methods and design of each. The aspects elucidated in this chapter are of vital importance, and provide a better understanding of power and energy engineering.

At the end, I would like to thank all those who dedicated their time and efforts for the successful completion of this book. I also wish to convey my gratitude towards my friends and family who supported me at every step.

Editor

1

Introduction to Power and Energy Engineering

Energy engineering deals with energy efficiency, environmental compliance, energy services and related fields. As a discipline it combines physics, math, chemistry and environmental engineering and is focused on increasing efficiency and developing alternate energy technologies. This introduction explores the topic of energy engineering and its subdivision, power engineering. This chapter is an overview of the subject matter incorporating all the major aspects of power and energy engineering.

Power Engineering

For the magazine, see Power Engineering (magazine). For the similar term but with a broad sense, see Energy engineering.

A steam turbine used to provide electric power.

Power engineering, also called power systems engineering, is a subfield of energy engineering and electrical engineering that deals with the generation, transmission, distribution and utilization of electric power and the electrical devices connected to such systems including generators, motors and transformers. Although much of the field is concerned with the problems of three-phase AC power – the standard for large-scale power transmission and distribution across the modern world – a significant fraction of the field is concerned with the conversion between AC and DC power and the development of specialized power systems such as those used in aircraft or for electric railway networks. Power engineering draws the majority of its theoretical base from electrical engineering and while some power engineers could be considered energy engineers, energy engineers often do not have the theoretical electrical engineering background to understand power engineering.

History

Electricity became a subject of scientific interest in the late 17th century with the work of William Gilbert. Over the next two centuries a number of important discoveries were made including the incandescent light bulb and the voltaic pile. Probably the greatest discovery with respect to power engineering came from Michael Faraday who in 1831 discovered that a change in magnetic flux induces an electromotive force in a loop of wire—a principle known as electromagnetic induction that helps explain how generators and transformers work.

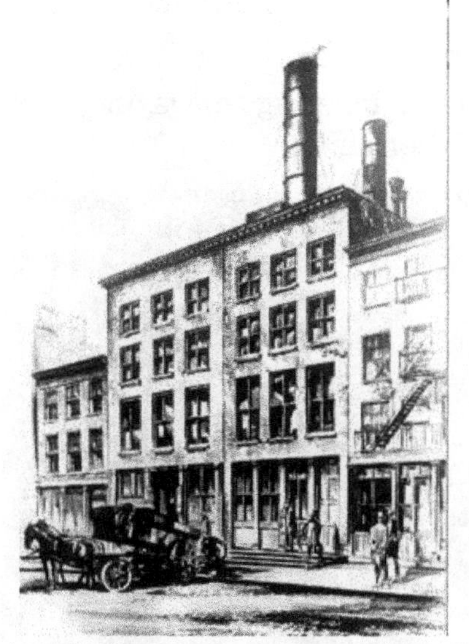

A sketch of the Pearl Street Station, the first steam-powered electric power station in New York City

In 1881 two electricians built the world's first power station at Godalming in England. The station employed two waterwheels to produce an alternating current that was used to supply seven Siemens arc lamps at 250 volts and thirty-four incandescent lamps at 40 volts. However supply was intermittent and in 1882 Thomas Edison and his company, The Edison Electric Light Company, developed the first steam-powered electric power station on Pearl Street in New York City. The Pearl Street Station consisted of several generators and initially powered around 3,000 lamps for 59 customers. The power station used direct current and operated at a single voltage. Since the direct current power could not be easily transformed to the higher voltages necessary to minimise power loss during transmission, the possible distance between the generators and load was limited to around half-a-mile (800 m).

That same year in London Lucien Gaulard and John Dixon Gibbs demonstrated the first transformer suitable for use in a real power system. The practical value of Gaulard and Gibbs' transformer was demonstrated in 1884 at Turin where the transformer was used to light up forty kilometres (25 miles) of railway from a single alternating current generator. Despite the success of the system, the pair made some fundamental mistakes. Perhaps the most serious was connecting the primaries of the transformers in series so that switching one lamp on or off would affect other lamps further down the line. Following the demonstration George Westinghouse, an American entrepreneur, imported a number of the transformers along with a Siemens generator and set his engineers to experimenting with them in the hopes of improving them for use in a commercial power system.

One of Westinghouse's engineers, William Stanley, recognised the problem with connecting transformers in series as opposed to parallel and also realised that making the iron core of a transformer a fully enclosed loop would improve the voltage regulation of the secondary winding. Using this knowledge he built the worlds first practical transformer based alternating current power system at Great Barrington, Massachusetts in 1886. In 1885 the Italian physicist and electrical engineer Galileo Ferraris demonstrated an induction motor and in 1887 and 1888 the Serbian-American engineer Nikola Tesla filed a range of patents related to power systems including one for a practical two-phase induction motor which Westinghouse licensed for his AC system.

By 1890 the power industry had flourished and power companies had built thousands of power systems (both direct and alternating current) in the United States and Europe – these networks were effectively dedicated to providing electric lighting. During this time a fierce rivalry in the US known as the "War of Currents" emerged between Edison and Westinghouse over which form of transmission (direct or alternating current) was superior. In 1891, Westinghouse installed the first major power system that was designed to drive an electric motor and not just provide electric lighting. The installation powered a 100 horsepower (75 kW) synchronous motor at Telluride, Colorado with the motor being started by a Tesla induction motor. On the other side of the Atlantic, Oskar von Miller built a 20 kV 176 km three-phase transmission line from Lauffen am Neckar to Frankfurt am Main for the Electrical Engineering Exhibition in Frankfurt. In 1895, after a protracted decision-making process, the Adams No. 1 generating station at Niagara Falls began transmitting three-phase alternating current power to Buffalo at 11 kV. Following completion of the Niagara Falls project, new power systems increasingly chose alternating current as opposed to direct current for electrical transmission.

Although the 1880s and 1890s were seminal decades in the field, developments in power engineering continued throughout the 20th and 21st century. In 1936 the first commercial high-voltage direct current (HVDC) line using mercury-arc valves was built between Schenectady and Mechanicville, New York. HVDC had previously been achieved by installing direct current generators in series (a system known as the Thury system) although this suffered from serious reliability issues. In 1957 Siemens demonstrated the first solid-state rectifier (solid-state rectifiers are now the standard for HVDC systems) however it was not until the early 1970s that this technology was used in commercial power systems. In 1959 Westinghouse demonstrated the first circuit breaker that used SF_6 as the interrupting medium. SF_6 is a far superior dielectric to air and, in recent times, its use has been extended to produce far more compact switching equipment (known as switchgear) and transformers. Many important developments also came from extending innovations in the ICT field to the power engineering field. For example, the development of computers meant load flow studies could be run more efficiently allowing for much better planning of power systems. Advances in information technology and telecommunication also allowed for much better remote control of the power system's switchgear and generators.

Basics of Electric Power

Electric power is the mathematical product of two quantities: current and voltage. These two quantities can vary with respect to time (AC power) or can be kept at constant levels (DC power).

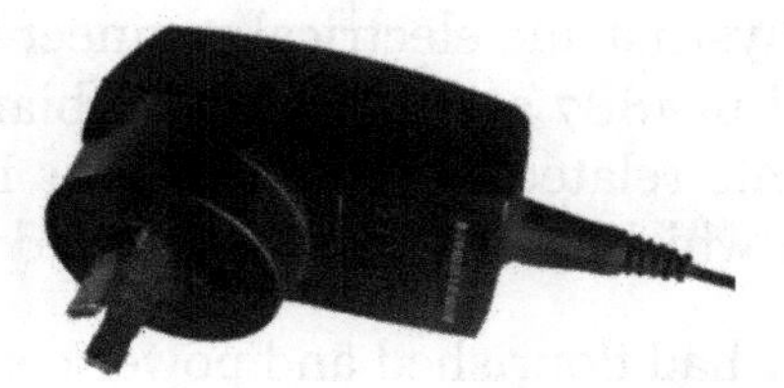

An external AC to DC power adapter used for household appliances

Most refrigerators, air conditioners, pumps and industrial machinery use AC power whereas computers and digital equipment use DC power (the digital devices you plug into the mains typically have an internal or external power adapter to convert from AC to DC power). AC power has the advantage of being easy to transform between voltages and is able to be generated and utilised by brushless machinery. DC power remains the only practical choice in digital systems and can be more economical to transmit over long distances at very high voltages.

The ability to easily transform the voltage of AC power is important for two reasons: Firstly, power can be transmitted over long distances with less loss at higher voltages. So in power networks where generation is distant from the load, it is desirable to step-up the voltage of power at the generation point and then step-down the voltage near the load. Secondly, it is often more economical to install turbines that produce higher voltages than would be used by most appliances, so the ability to easily transform voltages means this mismatch between voltages can be easily managed.

Solid state devices, which are products of the semiconductor revolution, make it possible to transform DC power to different voltages, build brushless DC machines and convert between AC and DC power. Nevertheless, devices utilising solid state technology are often more expensive than their traditional counterparts, so AC power remains in widespread use.

Power

Power Engineering deals with the generation, transmission, distribution and utilization of electricity as well as the design of a range of related devices. These include transformers, electric generators, electric motors and power electronics.

Transmission lines transmit power across the grid.

The power grid is an electrical network that connects a variety of electric generators to the users of electric power. Users purchase electricity from the grid so that they do not need to generate their own. Power engineers may work on the design and maintenance of the power grid as well as the power systems that connect to it. Such systems are called on-grid power systems and may supply the grid with additional power, draw power from the grid or do both. The grid is designed and managed using software that performs simulations of power flows.

Power engineers may also work on systems that do not connect to the grid. These systems are called off-grid power systems and may be used in preference to on-grid systems for a variety of reasons. For example, in remote locations it may be cheaper for a mine to generate its own power rather than pay for connection to the grid and in most mobile applications connection to the grid is simply not practical.

Today, most grids adopt three-phase electric power with alternating current. This choice can be partly attributed to the ease with which this type of power can be generated, transformed and used. Often (especially in the United States), the power is split before it reaches residential customers whose low-power appliances rely upon single-phase electric power. However, many larger industries and organizations still prefer to receive the three-phase power directly because it can be used to drive highly efficient electric motors such as three-phase induction motors.

Transformers play an important role in power transmission because they allow power to be converted to and from higher voltages. This is important because higher voltages suffer less power loss during transmission. This is because higher voltages allow for lower current to deliver the same amount of power, as power is the product of the two. Thus, as the voltage steps up, the current steps down. It is the current flowing through the components that result in both the losses and the subsequent heating. These losses, appearing in the form of heat, are equal to the current squared times the electrical resistance through which the current flows, so as the voltage goes up the losses are dramatically reduced.

For these reasons, electrical substations exist throughout power grids to convert power to higher voltages before transmission and to lower voltages suitable for appliances after transmission.

Components

Power engineering is a network of interconnected components which convert different forms of energy to electrical energy. Modern power engineering consists of four main subsystems: the generation subsystem, the transmission subsystem, the distribution subsystem and the utilization subsystem. In the generation subsystem, the power plant produces the electricity. The transmission subsystem transmits the electricity to the load centers. The distribution subsystem continues to transmit the power to the customers. The utilization system is concerned with the different uses of electrical energy like illumination, refrigeration, traction, electric drives, etc. Utilization is a very recent concept in Power engineering.

Generation

Generation of electrical power is a process whereby energy is transformed into an electrical form. There are several different transformation processes, among which are chemical, photo-voltaic, and electromechanical. Electromechanical energy conversion is used in converting energy from coal, petroleum, natural gas, uranium, or water flow into electrical energy. Of these, all except the wind energy conversion process take advantage of the synchronous AC generator coupled to a steam, gas or hydro turbine such that the turbine converts steam, gas, or water flow into rotational energy, and the synchronous generator then converts the rotational energy of the turbine into electrical energy. It is the turbine-generator conversion process that is by far most economical and consequently most common in the industry today.

The AC synchronous machine is the most common technology for generating electrical energy. It is called synchronous because the composite magnetic field produced by the three stator windings rotate at the same speed as the magnetic field produced by the field winding on the rotor. A simplified circuit model is used to analyze steady-state operating conditions for a synchronous machine. The phasor diagram is an effective tool for visualizing the relationships between internal voltage, armature current, and terminal voltage. The excitation control system is used on synchronous machines to regulate terminal voltage, and the turbine-governor system is used to regulate the speed of the machine. However, in highly interconnected systems, such as the "Western system", the "Texas system" and the "Eastern system", one machine will usually be assigned as the so-called "swing machine", and which generation may be increased or decreased to compensate for small changes in load, thereby maintaining the system frequency at precisely 60 Hz. Should the load dramatically change, as happens with a system separation, then a combination of "spinning reserve" and the "swing machine" may be used by the system's load dispatcher.

The operating costs of generating electrical energy is determined by the fuel cost and the efficiency of the power station. The efficiency depends on generation level and can be obtained from the heat rate curve. The incremental cost curve may also be derived from the heat rate curve. Economic dispatch is the process of allocating the required load demand between the available generation units such that the cost of operation is minimized. Emission dispatch is the process of allocating the required load demand between the available generation units such that air pollution occurring from operation is minimized. In large systems, particularly in the West, a combination of economic and emission dispatch may be used.

Transmission

The electricity is transported to load locations from a power station to a transmission subsystem. Therefore, we may think of the transmission system as providing the medium of transportation for electric energy. The transmission system may be subdivided into the bulk transmission system and the sub-transmission system. The functions of the bulk transmission are to interconnect generators, to interconnect various areas of the network, and to transfer electrical energy from the generators to the major load centers. This portion of the system is called "bulk" because it delivers energy only to so-called bulk loads such as the distribution system of a town, city, or large industrial plant. The function of the sub-transmission system is to interconnect the bulk power system with the distribution system.

Transmission circuits may be built either underground or overhead. Underground cables are used predominantly in urban areas where acquisition of overhead rights of way are costly or not possible. They are also used for transmission under rivers, lakes and bays. Overhead transmission is used otherwise because, for a given voltage level, overhead conductors are much less expensive than underground cables.

The transmission system is a highly integrated system. It is referred to as the substation equipment and transmission lines. The substation equipment contain the transformers, relays, and circuit breakers. Transformers are important static devices which transfer electrical energy from one circuit to another in the transmission subsystem. Transformers are used to step up the voltage on the transmission line to reduce the power loss which is dissipated on the way. A relay is functionally a level-detector; they perform a switching action when the input voltage (or current) meets or exceeds a specific and adjustable value. A circuit breaker is an automatically operated electrical switch designed to protect an electrical circuit from damage caused by overload or short circuit. A change in the status of any one component can significantly affect the operation of the entire system. Without adequate contact protection, the occurrence of undesired electric arcing causes significant degradation of the contacts, which suffer serious damage. There are three possible causes for power flow limitations to a transmission line. These causes are thermal overload, voltage instability, and rotor angle instability. Thermal overload is caused by excessive current flow in a circuit causing overheating. Voltage instability is said to occur when the power required to maintain voltages at or above acceptable levels exceeds the available power. Rotor angle instability is a dynamic problem that may occur following faults, such as short circuit, in the transmission system. It may also occur tens of seconds after a fault due to poorly damped or undamped oscillatory response of the rotor motion. As long as the equal area criteria is maintained, the interconnected system will remain stable. Should the equal area criteria be violated, it becomes necessary to separate the unstable component from the remainder of the system.

Distribution

The distribution system transports the power from the transmission system/substation to the customer. Distribution feeders can be radial or networked in an open loop configuration with a single or multiple alternate sources. Rural systems tend to be of the former and urban systems the latter. The equipment associated with the distribution system usually begins downstream of the distribution feeder circuit breaker. The transformer and circuit breaker are usually under the jurisdiction of a "substations department". The distribution feeders consist of combinations of overhead and underground conductor, 3 phase and single phase switches with load break and non-loadbreak ability, relayed protective devices, fuses, transformers (to utilization voltage), surge arresters, voltage regulators and capacitors.

More recently, Smart Grid initiatives are being deployed so that: One, distribution feeder faults are automatically isolated and power restored to unfaulted circuits by automatic hardware/software/communications packages. And two, capacitors are automatically switched on or off to dynamically control VAR flow and for CVR (Conservation Voltage Reduction).

Utilization

Utilization is the "end result" of the generation, transmission, and distribution of electric power. The energy carried by the transmission and distribution system is turned into useful work, light, heat, or a combination of these items at the utilization point. Understanding and characterizing the utilization of electric power is critical for proper planning and operation of power systems. Improper characterization of utilization can result of over or under building of power system facilities and stressing of system equipment beyond design capabilities. The term load refers to a device or collection of devices that draw energy from the power system. Individual loads (devices) range from small light bulbs to large induction motors to arc furnaces. The term load is often somewhat arbitrarily applied, at times being used to describe a specific device, and other times referring to an entire facility and even being used to describe the lumped power requirements of power system components and connected utilization devices downstream of a specific point in large scale system studies.

A major application of electric energy is in its conversion to mechanical energy. Electromagnetic, or "EM" devices designed for this purpose are commonly called "motors." Actually the machine is the central component of an integrated system consisting of the source, controller, motor, and load. For specialized applications, the system may be, and frequently is, designed as an integrated whole. Many household appliances (e.g., a vacuum cleaner) have in one unit, the controller, the motor, and the load.

Energy Engineering

Energy engineering or Energy systems is a broad field of engineering dealing with energy efficiency, energy services, facility management, plant engineering, environmental compliance and alternative energy technologies. Energy engineering is one of the more recent engineering disciplines to emerge. Energy engineering combines knowledge from the fields of physics, math, and chemistry with economic and environmental engineering practices. Energy engineers apply their skills to increase efficiency and further develop renewable sources of energy. The main job of energy engineers is to find the most efficient and sustainable ways to operate buildings and manufacturing processes. Energy engineers audit the use of energy in those processes and suggest ways to improve the systems. This means suggesting advanced lighting, better insulation, more efficient heating and cooling properties of buildings. Although an energy engineer is concerned about obtaining and using energy in the most environmentally friendly ways, their field is not limited to strictly renewable energy like hydro, solar, biomass, or geothermal. Energy engineers are also employed by the fields of oil and natural gas extraction.

Purpose

Energy minimization is the purpose of this growing discipline. Often applied to building design, heavy consideration is given to HVAC, lighting, refrigeration, to both reduce energy loads and increase efficiency of current systems. Energy engineering is increasingly seen as a major step forward in meeting carbon reduction targets. Since buildings and houses consume over 40% of the United States energy, the services an energy engineer performs are in demand.

History

Human beings have been transferring energy from one form to another since their use of fire. The efficiency of the transfer of energy is a new field. The oil crisis of 1973 and energy crisis of 1979 brought to light the need to get more work out of less energy. The United States government passed several laws in the seventies to promote increased energy efficiency, such as United States public law 94-413, the Federal Clean Car Incentive Program.

Power Engineering

Considered a subdivision of energy engineering, power engineering applies math and physics to the movement and transfer of energy to work in a system.

Leadership in Energy and Environmental Design

Leadership in Energy and Environmental Design (LEED) is a program created by the United States Green Building Council (USGBC) in March 2000. LEED is a program that encourages green building and promotes sustainability in the construction of buildings and the efficiency of the utilities in the buildings.

In 2012 the United States Green Building Council asked the independent firm Booz Allen Hamilton to conduct a study on the effectiveness of LEED program. "This study confirmed that green buildings generate substantial energy savings. From 2000–2008, green construction and renovation generated $1.3 billion in energy savings. Of that $1.3 billion, LEED-certified buildings accounted for $281 million." The study also found the summation of all green construction supported 2.4 million jobs.

Energy Efficiency

Energy efficiency is seen two ways. The first view is that more work is done from the same amount of energy used. The other perception is that the same amount of work is accomplished with less energy used in the system. Some ways to get more work out of less energy is to "Reduce, Reuse, and Recycle" the materials used in daily life. The advancement of technology has led to other uses of waste. Technology such as waste-to-energy facilities which convert solid wastes through the process of gasification or pyrolysis to liquid fuels to be burned. The Environmental Protection Agency stated that the United States produced 250 million tons of municipal waste in 2010. Of that 250 million tons roughly 54% gets thrown in land fills, 33% is recycled, and 13% goes to energy recovery plants. In European countries who pay more for fuel, such as Denmark where the price for a gallon of gas neared $10 in 2010, have more fully developed waste-to energy facilities. In 2010 Denmark sent 7% of waste to landfills, 69% was recycled, and 24% was sent to waste-to-energy facilities. There are several other developed Western European countries that also have taken energy engineering into consideration. Germany's "Energiewende", a policy which set the goal by 2050 to meet 80% of electrical needs from renewable energy sources.

Statistics

The median yearly salary for an energy engineering is $64,587 U.S. dollars. 83% of energy engineers are male while the remaining 17% are female. 65% of energy engineers have less than five years of experience in their profession.

Professional Organizations

Energy engineers have one main professional organization. The Association of Energy Engineers (AEE) founded in 1977. It now has 16,000 members in 89 countries. The organizations main responsibility is to lead energy certification programs to teach individuals across the world and certify as qualified to perform the job of an energy engineer.

Education

A bachelor's degree is a primary requirement of becoming an energy engineer. Some energy engineers are registered Professional Engineers (P.E.s); although the completion of that program is not a necessity. Of the 16,000 energy engineers that are in the Association of Energy Engineers (AEE), 58.4% have post-graduate degrees from accredited colleges. If the engineer is not a P.E. he or she would need a Certified Engineering Manager (CEM) certificate from a group like the AEE.

While a student who wants to become an energy engineer does not directly need to get a degree in energy engineering, several universities across the world have established departments or centers offering energy engineering degrees, to better prepare future engineers for their career.

Electrical Engineering

Electrical engineering is a field of engineering that generally deals with the study and application of electricity, electronics, and electromagnetism. This field first became an identifiable occupation in the latter half of the 19th century after commercialization of the electric telegraph, the telephone, and electric power distribution and use. Subsequently, broadcasting and recording media made electronics part of daily life. The invention of the transistor, and later the integrated circuit, brought down the cost of electronics to the point they can be used in almost any household object.

Electrical engineers design complex power systems ...

Electrical engineering has now subdivided into a wide range of subfields including electronics, digital computers, power engineering, telecommunications, control systems, radio-frequency engineering, signal processing, instrumentation, and microelectronics. The subject of electronic engineering is often treated as its own subfield but it intersects with all the other subfields, including the power electronics of power engineering.

... and electronic circuits.

Electrical engineers typically hold a degree in electrical engineering or electronic engineering. Practicing engineers may have professional certification and be members of a professional body. Such bodies include the Institute of Electrical and Electronics Engineers (IEEE) and the Institution of Engineering and Technology (professional society) (IET).

Electrical engineers work in a very wide range of industries and the skills required are likewise variable. These range from basic circuit theory to the management skills required of a project manager. The tools and equipment that an individual engineer may need are similarly variable, ranging from a simple voltmeter to a top end analyzer to sophisticated design and manufacturing software.

History

Electricity has been a subject of scientific interest since at least the early 17th century. A prominent early electrical scientist was William Gilbert who was the first to draw a clear distinction between magnetism and static electricity and is credited with establishing the term electricity. He also designed the versorium: a device that detected the presence of statically charged objects. Then in 1762 Swedish professor Johan Carl Wilcke invented, and in 1775 Alessandro Volta improved, a device (for which Volta coined the name electrophorus) that produced a static electric charge, and by 1800 Volta had developed the voltaic pile, a forerunner of the electric battery.

19th Century

In the 19th century, research into the subject started to intensify. Notable developments in this century include the work of Georg Ohm, who in 1827 quantified the relationship between the electric current and potential difference in a conductor, of Michael Faraday, the discoverer of electromagnetic induction in 1831, and of James Clerk Maxwell, who in 1873 published a unified theory of electricity and magnetism in his treatise Electricity and Magnetism.

The discoveries of Michael Faraday formed the foundation of electric motor technology

Electrical engineering became a profession in the later 19th century. Practitioners had created a global electric telegraph network and the first professional electrical engineering institutions were founded in the UK and USA to support the new discipline. Although it is impossible to precisely pinpoint a first electrical engineer, Francis Ronalds stands ahead of the field, who created the first working electric telegraph system in 1816 and documented his vision of how the world could be transformed by electricity. Over 50 years later, he joined the new Society of Telegraph Engineers (soon to be renamed the Institution of Electrical Engineers) where he was regarded by other members as the first of their cohort. By the end of the 19th century, the world had been forever changed by the rapid communication made possible by the engineering development of landlines, submarine cables, and, from about 1890, wireless telegraphy.

Practical applications and advances in such fields created an increasing need for standardised units of measure. They led to the international standardization of the units volt, ampere, coulomb, ohm, farad, and henry. This was achieved at an international conference in Chicago in 1893. The publication of these standards formed the basis of future advances in standardisation in various industries, and in many countries the definitions were immediately recognised in relevant legislation.

During these years, the study of electricity was largely considered to be a subfield of physics. That's because early electrical technology was electromechanical in nature. The Technische Universität Darmstadt founded the world's first department of electrical engineering in 1882. The first electrical engineering degree program was started at Massachusetts Institute of Technology (MIT) in the physics department under Professor Charles Cross, though it was Cornell University to produce the world's first electrical engineering graduates in 1885. The first course in electrical engineering was taught in 1883 in Cornell's Sibley College of Mechanical Engineering and Mechanic Arts. It was not until about 1885 that Cornell President Andrew Dickson White established the first Department of Electrical Engineering in the United States. In the same year, University College London founded the first chair of electrical engineering in Great Britain. Professor Mendell P. Weinbach at University of Missouri soon followed suit by establishing the electrical engineering department in 1886. Afterwards, universities and institutes of technology gradually started to offer electrical engineering programs to their students all over the world.

During these decades use of electrical engineering increased dramatically. In 1882, Thomas Edison switched on the world's first large-scale electric power network that provided 110 volts — direct current (DC) — to 59 customers on Manhattan Island in New York City. In 1884, Sir Charles Parsons invented the steam turbine allowing for more efficient electric power generation. Alternating current, with its ability to transmit power more efficiently over long distances via the use of transformers, developed rapidly in the 1880s and 1890s with transformer designs by Károly Zipernowsky, Ottó Bláthy and Miksa Déri (later called ZBD transformers), Lucien Gaulard, John Dixon Gibbs and William Stanley, Jr.. Practical AC motor designs including induction motors were independently invented by Galileo Ferraris and Nikola Tesla and further developed into a practical three-phase form by Mikhail Dolivo-Dobrovolsky and Charles Eugene Lancelot Brown. Charles Steinmetz and Oliver Heaviside contributed to the theoretical basis of alternating current engineering. The spread in the use of AC set off in the United States what has been called the War of Currents between a George Westinghouse backed AC system and a Thomas Edison backed DC power system, with AC being adopted as the overall standard.

More Modern Developments

During the development of radio, many scientists and inventors contributed to radio technology and electronics. The mathematical work of James Clerk Maxwell during the 1850s had shown the relationship of different forms of electromagnetic radiation including possibility of invisible airborne waves (later called "radio waves"). In his classic physics experiments of 1888, Heinrich Hertz proved Maxwell's theory by transmitting radio waves with a spark-gap transmitter, and detected them by using simple electrical devices. Other physicists experimented with these new waves and in the process developed devices for transmitting and detecting them. In 1895, Guglielmo Marconi began work on a way to adapt the known methods of transmitting and detecting these "Hertzian waves" into a purpose built commercial wireless telegraphic system. Early on, he sent wireless signals over a distance of one and a half miles. In December 1901, he sent wireless waves that were not affected by the curvature of the Earth. Marconi later transmitted the wireless signals across the Atlantic between Poldhu, Cornwall, and St. John's, Newfoundland, a distance of 2,100 miles (3,400 km).

Guglielmo Marconi known for his pioneering work on long distance radio transmission

In 1897, Karl Ferdinand Braun introduced the cathode ray tube as part of an oscilloscope, a crucial enabling technology for electronic television. John Fleming invented the first radio tube, the diode, in 1904. Two years later, Robert von Lieben and Lee De Forest independently developed the amplifier tube, called the triode.

In 1920, Albert Hull developed the magnetron which would eventually lead to the development of the microwave oven in 1946 by Percy Spencer. In 1934, the British military began to make strides toward radar (which also uses the magnetron) under the direction of Dr Wimperis, culminating in the operation of the first radar station at Bawdsey in August 1936.

A replica of the first working transistor.

In 1941, Konrad Zuse presented the Z3, the world's first fully functional and programmable computer using electromechanical parts. In 1943, Tommy Flowers designed and built the Colossus, the world's first fully functional, electronic, digital and programmable computer. In 1946, the ENIAC (Electronic Numerical Integrator and Computer) of John Presper Eckert and John Mauchly followed, beginning the computing era. The arithmetic performance of these machines allowed engineers to develop completely new technologies and achieve new objectives, including the Apollo program which culminated in landing astronauts on the Moon.

Solid-state Transistors

The invention of the transistor in late 1947 by William B. Shockley, John Bardeen, and Walter Brattain of the Bell Telephone Laboratories opened the door for more compact devices and led to the development of the integrated circuit in 1958 by Jack Kilby and independently in 1959 by Robert Noyce. Starting in 1968, Ted Hoff and a team at the Intel Corporation invented the first commercial microprocessor, which foreshadowed the personal computer. The Intel 4004 was a four-bit processor released in 1971, but in 1973 the Intel 8080, an eight-bit processor, made the first personal computer, the Altair 8800, possible.

Subdisciplines

Electrical engineering has many subdisciplines, the most common of which are listed below. Although there are electrical engineers who focus exclusively on one of these subdisciplines, many deal with a combination of them. Sometimes certain fields, such as electronic engineering and computer engineering, are considered separate disciplines in their own right.

Power

Power pole

Power engineering deals with the generation, transmission, and distribution of electricity as well as the design of a range of related devices. These include transformers, electric generators, electric motors, high voltage engineering, and power electronics. In many regions of the world, governments maintain an electrical network called a power grid that connects a variety of generators together with users of their energy. Users purchase electrical energy from the grid, avoiding the costly exercise of having to generate their own. Power engineers may work on the design and maintenance of the power grid as well as the power systems that connect to it. Such systems are called on-grid power systems and may supply the grid with additional power, draw power from the grid, or do both. Power engineers may also work on systems that do not connect to the grid, called off-grid power systems, which in some cases are preferable to on-grid systems. The future includes Satellite controlled power systems, with feedback in real time to prevent power surges and prevent blackouts.

Control

Control engineering focuses on the modeling of a diverse range of dynamic systems and the design of controllers that will cause these systems to behave in the desired manner. To implement such controllers, electrical engineers may use electronic circuits, digital signal processors, microcontrollers, and programmable logic controls (PLCs). Control engineering has a wide range of applications from the flight and propulsion systems of commercial airliners to the cruise control present in many modern automobiles. It also plays an important role in industrial automation.

Control systems play a critical role in space flight.

Control engineers often utilize feedback when designing control systems. For example, in an automobile with cruise control the vehicle's speed is continuously monitored and fed back to the system which adjusts the motor's power output accordingly. Where there is regular feedback, control theory can be used to determine how the system responds to such feedback.

Electronics

Electronic engineering involves the design and testing of electronic circuits that use the properties of components such as resistors, capacitors, inductors, diodes, and transistors to achieve a particular functionality. The tuned circuit, which allows the user of a radio to filter out all but a single station, is just one example of such a circuit. Another example (of a pneumatic signal conditioner) is shown in the adjacent photograph.

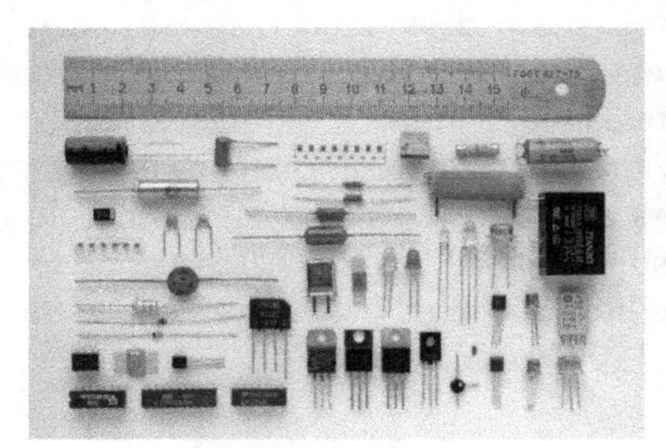

Electronic components

Prior to the Second World War, the subject was commonly known as radio engineering and basically was restricted to aspects of communications and radar, commercial radio, and early television. Later, in post war years, as consumer devices began to be developed, the field grew to include modern television, audio systems, computers, and microprocessors. In the mid-to-late 1950s, the term radio engineering gradually gave way to the name electronic engineering.

Before the invention of the integrated circuit in 1959, electronic circuits were constructed from discrete components that could be manipulated by humans. These discrete circuits consumed much space and power and were limited in speed, although they are still common in some applications. By contrast, integrated circuits packed a large number—often millions—of tiny electrical components, mainly transistors, into a small chip around the size of a coin. This allowed for the powerful computers and other electronic devices we see today.

Microelectronics

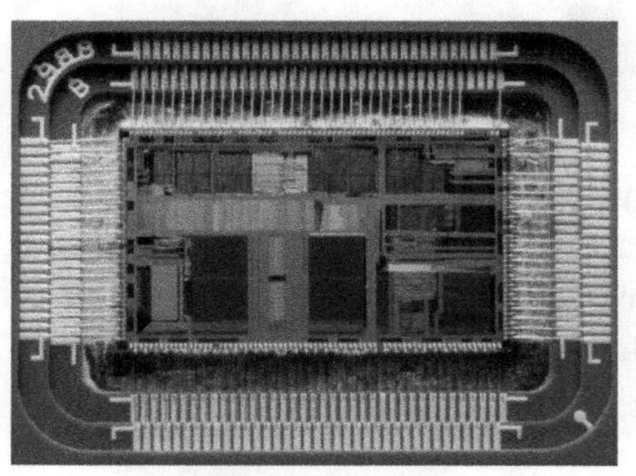

Microprocessor

Microelectronics engineering deals with the design and microfabrication of very small electronic circuit components for use in an integrated circuit or sometimes for use on their own as a general electronic component. The most common microelectronic components are semiconductor transistors, although all main electronic components (resistors, capacitors etc.) can be created at a microscopic level. Nanoelectronics is the further scaling of devices down to nanometer levels. Modern devices are already in the nanometer regime, with below 100 nm processing having been standard since about 2002.

Microelectronic components are created by chemically fabricating wafers of semiconductors such as silicon (at higher frequencies, compound semiconductors like gallium arsenide and indium phosphide) to obtain the desired transport of electronic charge and control of current. The field of microelectronics involves a significant amount of chemistry and material science and requires the electronic engineer working in the field to have a very good working knowledge of the effects of quantum mechanics.

Signal Processing

Signal processing deals with the analysis and manipulation of signals. Signals can be either analog, in which case the signal varies continuously according to the information, or digital, in which case the signal varies according to a series of discrete values representing the information. For analog signals, signal processing may involve the amplification and filtering of audio signals for audio equipment or the modulation and demodulation of signals for telecommunications. For digital signals, signal processing may involve the compression, error detection and error correction of digitally sampled signals.

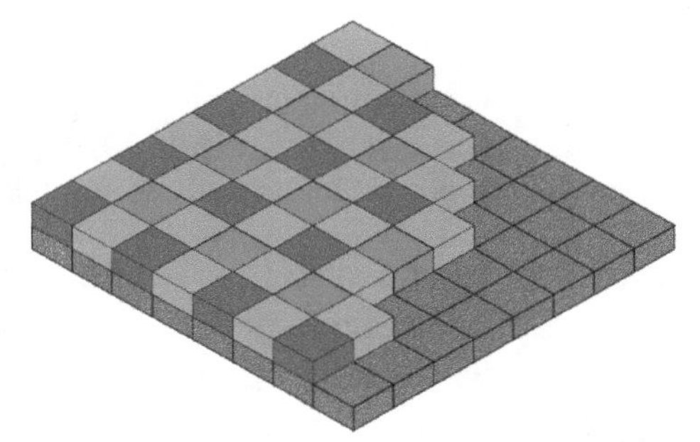

A Bayer filter on a CCD requires signal processing to get a red, green, and blue value at each pixel.

Signal Processing is a very mathematically oriented and intensive area forming the core of digital signal processing and it is rapidly expanding with new applications in every field of electrical engineering such as communications, control, radar, audio engineering, broadcast engineering, power electronics, and biomedical engineering as many already existing analog systems are replaced with their digital counterparts. Analog signal processing is still important in the design of many control systems.

DSP processor ICs are found in every type of modern electronic systems and products including, SDTV | HDTV sets, radios and mobile communication devices, Hi-Fi audio equipment, Dolby noise reduction algorithms, GSM mobile phones, mp3 multimedia players, camcorders and digital cameras, automobile control systems, noise cancelling headphones, digital spectrum analyzers, intelligent missile guidance, radar, GPS based cruise control systems, and all kinds of image processing, video processing, audio processing, and speech processing systems.

Telecommunications

Telecommunications engineering focuses on the transmission of information across a channel such as a coax cable, optical fiber or free space. Transmissions across free space require information to be encoded in a carrier signal to shift the information to a carrier frequency suitable for transmission; this is known as modulation. Popular analog modulation techniques include amplitude modulation and frequency modulation. The choice of modulation affects the cost and performance of a system and these two factors must be balanced carefully by the engineer.

Satellite dishes are a crucial component in the analysis of satellite information.

Once the transmission characteristics of a system are determined, telecommunication engineers design the transmitters and receivers needed for such systems. These two are sometimes combined to form a two-way communication device known as a transceiver. A key consideration in the design of transmitters is their power consumption as this is closely related to their signal strength. If the signal strength of a transmitter is insufficient the signal's information will be corrupted by noise.

Instrumentation

Instrumentation engineering deals with the design of devices to measure physical quantities such as pressure, flow, and temperature. The design of such instrumentation requires a good understanding of physics that often extends beyond electromagnetic theory. For example, flight instruments measure variables such as wind speed and altitude to enable pilots the control of aircraft analytically. Similarly, thermocouples use the Peltier-Seebeck effect to measure the temperature difference between two points.

Flight instruments provide pilots with the tools to control aircraft analytically.

Often instrumentation is not used by itself, but instead as the sensors of larger electrical systems. For example, a thermocouple might be used to help ensure a furnace's temperature remains constant. For this reason, instrumentation engineering is often viewed as the counterpart of control engineering.

Computers

Computer engineering deals with the design of computers and computer systems. This may involve the design of new hardware, the design of PDAs, tablets, and supercomputers, or the use of computers to control an industrial plant. Computer engineers may also work on a system's software. However, the design of complex software systems is often the domain of software engineering, which is usually considered a separate discipline. Desktop computers represent a tiny fraction of the devices a computer engineer might work on, as computer-like architectures are now found in a range of devices including video game consoles and DVD players.

Supercomputers are used in fields as diverse as computational biology and geographic information systems.

Related Disciplines

Mechatronics is an engineering discipline which deals with the convergence of electrical and mechanical systems. Such combined systems are known as electromechanical systems and have widespread adoption. Examples include automated manufacturing systems, heating, ventilation and air-conditioning systems, and various subsystems of aircraft and automobiles.

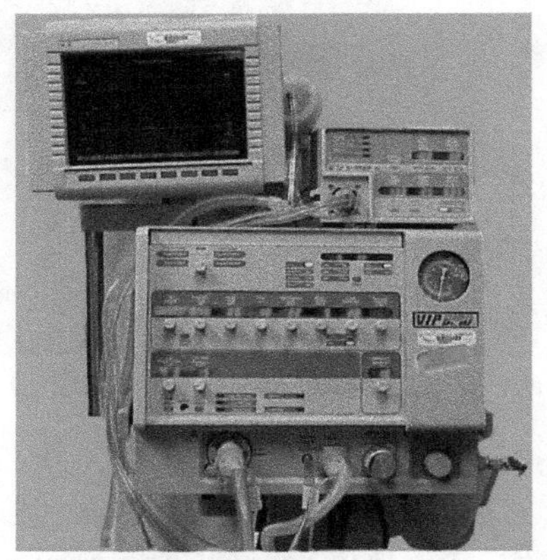

The Bird VIP Infant ventilator

The term mechatronics is typically used to refer to macroscopic systems but futurists have predicted the emergence of very small electromechanical devices. Already, such small devices, known as Microelectromechanical systems (MEMS), are used in automobiles to tell airbags when to deploy, in digital projectors to create sharper images, and in inkjet printers to create nozzles for high definition printing. In the future it is hoped the devices will help build tiny implantable medical devices and improve optical communication.

Biomedical engineering is another related discipline, concerned with the design of medical equipment. This includes fixed equipment such as ventilators, MRI scanners, and electrocardiograph monitors as well as mobile equipment such as cochlear implants, artificial pacemakers, and artificial hearts.

Aerospace engineering and robotics an example is the most recent electric propulsion and ion propulsion.

Education

Electrical engineers typically possess an academic degree with a major in electrical engineering, electronics engineering, electrical engineering technology, or electrical and electronic engineering. The same fundamental principles are taught in all programs, though emphasis may vary according to title. The length of study for such a degree is usually four or five years and the completed degree may be designated as a Bachelor of Science in Electrical/Electronics Engineering Technology, Bachelor of Engineering, Bachelor of Science, Bachelor of Technology, or Bachelor of Applied Science depending on the university. The bachelor's degree generally includes units covering physics, mathematics, computer science, project management, and a variety of topics in electrical engineering. Initially such topics cover most, if not all, of the subdisciplines of electrical engineering. At some schools, the students can then choose to emphasize one or more subdisciplines towards the end of their courses of study.

Oscilloscope

At many schools, electronic engineering is included as part of an electrical award, sometimes explicitly, such as a Bachelor of Engineering (Electrical and Electronic), but in others electrical and electronic engineering are both considered to be sufficiently broad and complex that separate degrees are offered.

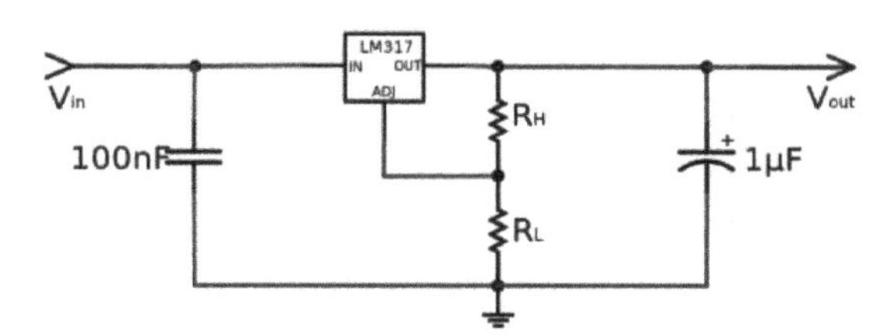

Typical electrical engineering diagram used as a troubleshooting tool

Some electrical engineers choose to study for a postgraduate degree such as a Master of Engineering/Master of Science (M.Eng./M.Sc.), a Master of Engineering Management, a Doctor of Philosophy (Ph.D.) in Engineering, an Engineering Doctorate (Eng.D.), or an Engineer's degree. The master's and engineer's degrees may consist of either research, coursework or a mixture of the two. The Doctor of Philosophy and Engineering Doctorate degrees consist of a significant research component and are often viewed as the entry point to academia. In the United Kingdom and some other European countries, Master of Engineering is often considered to be an undergraduate degree of slightly longer duration than the Bachelor of Engineering rather than postgraduate.

Practicing Engineers

Belgian electrical engineers inspecting the rotor of a 40,000 kilowatt turbine of the General Electric Company in New York City

In most countries, a bachelor's degree in engineering represents the first step towards professional certification and the degree program itself is certified by a professional body. After completing a certified degree program the engineer must satisfy a range of requirements (including work experience requirements) before being certified. Once certified the engineer is designated the title of Professional Engineer (in the United States, Canada and South Africa), Chartered Engineer or Incorporated Engineer (in India, Pakistan, the United Kingdom, Ireland and Zimbabwe), Chartered Professional Engineer (in Australia and New Zealand) or European Engineer (in much of the European Union).

The IEEE corporate office is on the 17th floor of 3 Park Avenue in New York City

The advantages of certification vary depending upon location. For example, in the United States and Canada "only a licensed engineer may seal engineering work for public and private clients". This requirement is enforced by state and provincial legislation such as Quebec's Engineers Act. In other countries, no such legislation exists. Practically all certifying bodies maintain a code of ethics that they expect all members to abide by or risk expulsion. In this way these organizations play an important role in maintaining ethical standards for the profession. Even in jurisdictions where certification has little or no legal bearing on work, engineers are subject to contract law. In cases where an engineer's work fails he or she may be subject to the tort of negligence and, in extreme cases, the charge of criminal negligence. An engineer's work must also comply with numerous other rules and regulations such as building codes and legislation pertaining to environmental law.

Professional bodies of note for electrical engineers include the Institute of Electrical and Electronics Engineers (IEEE) and the Institution of Engineering and Technology (IET). The IEEE claims to produce 30% of the world's literature in electrical engineering, has over 360,000 members worldwide and holds over 3,000 conferences annually. The IET publishes 21 journals, has a worldwide membership of over 150,000, and claims to be the largest professional engineering society in Europe. Obsolescence of technical skills is a serious concern for electrical engineers. Membership and participation in technical societies, regular reviews of periodicals in the field and a habit of continued learning are therefore essential to maintaining proficiency. An MIET(Member of the Institution of Engineering and Technology) is recognised in Europe as an Electrical and computer (technology) engineer.

In Australia, Canada, and the United States electrical engineers make up around 0.25% of the labor force.

Tools and Work

From the Global Positioning System to electric power generation, electrical engineers have contributed to the development of a wide range of technologies. They design, develop, test, and supervise the deployment of electrical systems and electronic devices. For example, they may work on the design of telecommunication systems, the operation of electric power stations, the lighting and wiring of buildings, the design of household appliances, or the electrical control of industrial machinery.

Satellite communications is typical of what electrical engineers work on.

Fundamental to the discipline are the sciences of physics and mathematics as these help to obtain both a qualitative and quantitative description of how such systems will work. Today most engineering work involves the use of computers and it is commonplace to use computer-aided design programs when designing electrical systems. Nevertheless, the ability to sketch ideas is still invaluable for quickly communicating with others.

The Shadow robot hand system

Although most electrical engineers will understand basic circuit theory (that is the interactions of elements such as resistors, capacitors, diodes, transistors, and inductors in a circuit), the theories employed by engineers generally depend upon the work they do. For example, quantum mechanics and solid state physics might be relevant to an engineer working on VLSI (the design of integrated circuits), but are largely irrelevant to engineers working with macroscopic electrical systems. Even circuit theory may not be relevant to a person designing telecommunication systems that use off-the-shelf components. Perhaps the most important technical skills for electrical engineers are reflected in university programs, which emphasize strong numerical skills, computer literacy, and the ability to understand the technical language and concepts that relate to electrical engineering.

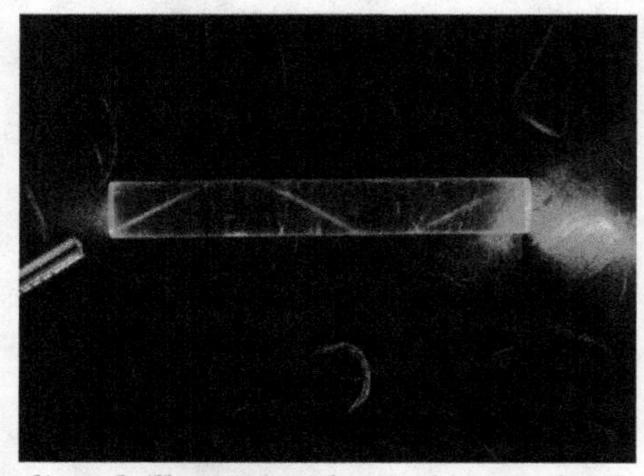

A laser bouncing down an acrylic rod, illustrating the total internal reflection of light in a multi-mode optical fiber.

A wide range of instrumentation is used by electrical engineers. For simple control circuits and alarms, a basic multimeter measuring voltage, current, and resistance may suffice. Where time-varying signals need to be studied, the oscilloscope is also an ubiquitous instrument. In RF engineering and high frequency telecommunications, spectrum analyzers and network analyzers are used. In some disciplines, safety can be a particular concern with instrumentation. For instance, medical electronics designers must take into account that much lower voltages than normal can be dangerous when electrodes are directly in contact with internal body fluids. Power transmission engineering also has great safety concerns due to the high voltages used; although voltmeters may in principle be similar to their low voltage equivalents, safety and calibration issues make them very different. Many disciplines of electrical engineering use tests specific to their discipline. Audio electronics engineers use audio test sets consisting of a signal generator and a meter, principally to measure level but also other parameters such as harmonic distortion and noise. Likewise, information technology have their own test sets, often specific to a particular data format, and the same is true of television broadcasting.

For many engineers, technical work accounts for only a fraction of the work they do. A lot of time may also be spent on tasks such as discussing proposals with clients, preparing budgets and determining project schedules. Many senior engineers manage a team of technicians or other engineers and for this reason project management skills are important. Most engineering projects involve some form of documentation and strong written communication skills are therefore very important.

Radome at the Misawa Air Base Misawa Security Operations Center, Misawa, Japan

The workplaces of engineers are just as varied as the types of work they do. Electrical engineers may be found in the pristine lab environment of a fabrication plant, the offices of a consulting firm or on site at a mine. During their working life, electrical engineers may find themselves supervising a wide range of individuals including scientists, electricians, computer programmers, and other engineers.

Electrical engineering has an intimate relationship with the physical sciences. For instance, the physicist Lord Kelvin played a major role in the engineering of the first transatlantic telegraph cable. Conversely, the engineer Oliver Heaviside produced major work on the mathematics of transmission on telegraph cables. Electrical engineers are often required on major science projects. For instance, large particle accelerators such as CERN need electrical engineers to deal with many aspects of the project: from the power distribution, to the instrumentation, to the manufacture and installation of the superconducting electromagnets.

References

- Ned Mohan, T. M. Undeland and William P. Robbins (2003). Power Electronics: Converters, Applications, and Design. United States of America: John Wiley & Sons, Inc. ISBN 0-471-22693-9.
- Ronalds, B.F. (2016). Sir Francis Ronalds: Father of the Electric Telegraph. London: Imperial College Press. ISBN 978-1-78326-917-4.
- Engineering: Issues, Challenges and Opportunities for Development. UNESCO. 2010. pp. 127–8. ISBN 978-92-3-104156-3.
- Manual on the Use of Thermocouples in Temperature Measurement. ASTM International. 1 January 1993. p. 154. ISBN 978-0-8031-1466-1.
- Occupational Outlook Handbook, 2008–2009. U S Department of Labor, Jist Works. 1 March 2008. p. 148. ISBN 978-1-59357-513-7.
- “Electrical and Electronic Engineer”. Occupational Outlook Handbook, 2012-13 Edition. Bureau of Labor Statistics, U.S. Department of Labor. Retrieved November 15, 2014.

- Baake, Rainer; Morgan, Jennifer (May 15, 2013). "U.S. Energy Policy Should Take a Lesson From Germany's Energiewende". Bloomberg. Retrieved 2013-06-13.
- Battles, Stephanie J.; Burns, Eugene M. (October 17, 1999). "United States Energy Usage and Efficiency: Measuring Changes Over Time". Energy Information Administration. Retrieved 2013-06-11.
- "What is the difference between electrical and electronic engineering?". FAQs - Studying Electrical Engineering. Retrieved 20 March 2012.
- Fritz E. Froehlich, Allen Kent, The Froehlich/Kent Encyclopedia of Telecommunications: Volume 17, page 36. Books.google.com. Retrieved 2012-09-10.
- "Lab Note #105 Contact Life – Unsuppressed vs. Suppressed Arcing". Arc Suppression Technologies. April 2011. Retrieved February 5, 2012.

2

Power Electronics: Features and Applications

Power electronics is concerned with the conversion of power and the devices using it may be employed as switches or amplifiers, which help in power modulation and regulation. The chapter studies devices like DC- to-DC converter, rectifier and power inverter. The attributes and features of each have been discussed in detail to enable an insightful understanding of the complex subject-matter.

Power Electronics

An HVDC thyristor valve tower 16.8 m tall in a hall at Baltic Cable AB in Sweden

Power electronics is the application of solid-state electronics to the control and conversion of electric power. It also refers to a subject of research in electronic and electrical engineering which deals with the design, control, computation and integration of nonlinear, time-varying energy-processing electronic systems with fast dynamics.

The first high power electronic devices were mercury-arc valves. In modern systems the conversion is performed with semiconductor switching devices such as diodes, thyristors and transistors, pioneered by R. D. Middlebrook and others beginning in the 1950s. In contrast to electronic systems concerned with transmission and processing of signals and data, in power electronics substantial amounts of electrical energy are processed. An AC/DC converter (rectifier) is the most typical power electronics device found in many consumer electronic devices, e.g. television sets, personal computers, battery chargers, etc. The power range is typically from tens of watts to several hundred watts. In industry a common application is the variable speed drive (VSD) that is used to control an induction motor. The power range of VSDs start from a few hundred watts and end at tens of megawatts.

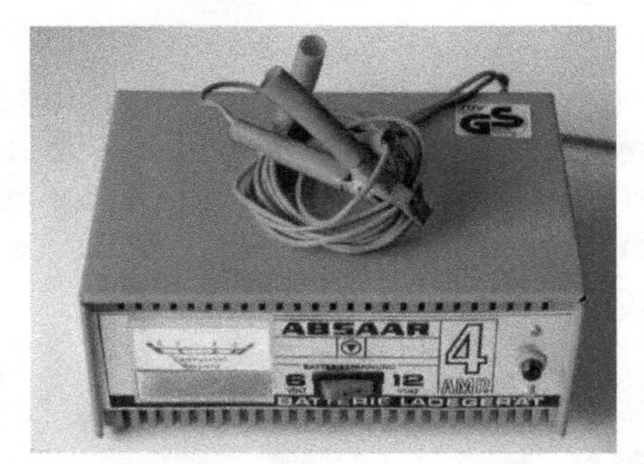

A battery charger is an example of a piece of power electronics

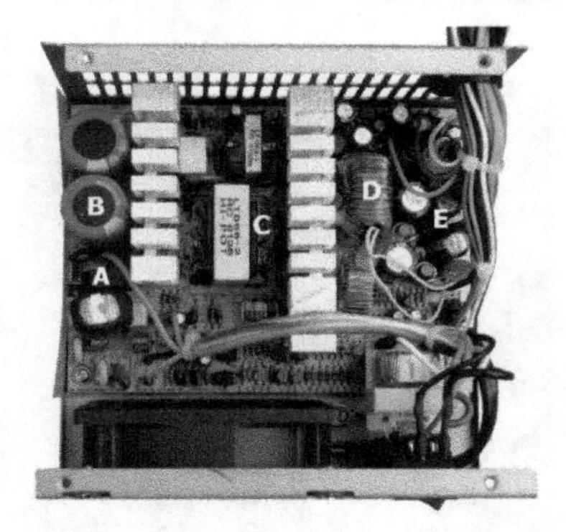

A PCs power supply is an example of a piece of power electronics, whether inside or outside of the cabinet

The power conversion systems can be classified according to the type of the input and output power

- AC to DC (rectifier)
- DC to AC (inverter)
- DC to DC (DC-to-DC converter)
- AC to AC (AC-to-AC converter)

History

Power electronics started with the development of the mercury arc rectifier. Invented by Peter Cooper Hewitt in 1902, it was used to convert alternating current (AC) into direct current (DC). From the 1920s on, research continued on applying thyratrons and grid-controlled mercury arc valves to power transmission. Uno Lamm developed a mercury valve with grading electrodes making them suitable for high voltage direct current power transmission. In 1933 selenium rectifiers were invented.

In 1947 the bipolar point-contact transistor was invented by Walter H. Brattain and John Bardeen under the direction of William Shockley at Bell Labs. In 1948 Shockley's invention of the bipolar junction transistor (BJT) improved the stability and performance of transistors, and reduced costs. By the 1950s, higher power semiconductor diodes became available and started replacing vacuum tubes. In 1956 the silicon controlled rectifier (SCR) was introduced by General Electric, greatly increasing the range of power electronics applications.

By the 1960s the improved switching speed of bipolar junction transistors had allowed for high frequency DC/DC converters. In 1976 power MOSFETs became commercially available. In 1982 the Insulated Gate Bipolar Transistor (IGBT) was introduced.

Devices

The capabilities and economy of power electronics system are determined by the active devices that are available. Their characteristics and limitations are a key element in the design of power electronics systems. Formerly, the mercury arc valve, the high-vacuum and gas-filled diode thermionic rectifiers, and triggered devices such as the thyratron and ignitron were widely used in power electronics. As the ratings of solid-state devices improved in both voltage and current-handling capacity, vacuum devices have been nearly entirely replaced by solid-state devices.

Power electronic devices may be used as switches, or as amplifiers. An ideal switch is either open or closed and so dissipates no power; it withstands an applied voltage and passes no current, or passes any amount of current with no voltage drop. Semiconductor devices used as switches can approximate this ideal property and so most power electronic applications rely on switching devices on and off, which makes systems very efficient as very little power is wasted in the switch. By contrast, in the case of the amplifier, the current through the device varies continuously according to a controlled input. The voltage and current at the device terminals follow a load line, and the power dissipation inside the device is large compared with the power delivered to the load.

Several attributes dictate how devices are used. Devices such as diodes conduct when a forward voltage is applied and have no external control of the start of conduction. Power devices such as silicon controlled rectifiers and thyristors (as well as the mercury valve and thyratron) allow control of the start of conduction, but rely on periodic reversal of current flow to turn them off. Devices such as gate turn-off thyristors, BJT and MOSFET transistors provide full switching control and can be turned on or off without regard to the current flow through them. Transistor devices also allow proportional amplification, but this is rarely used for systems rated more than a few hundred watts. The control input characteristics of a device also greatly affect design; sometimes the control input is at a very high voltage with respect to ground and must be driven by an isolated source.

As efficiency is at a premium in a power electronic converter, the losses that a power electronic device generates should be as low as possible.

Devices vary in switching speed. Some diodes and thyristors are suited for relatively slow speed and are useful for power frequency switching and control; certain thyristors are useful at a few kilohertz. Devices such as MOSFETS and BJTs can switch at tens of kilohertz up to a few megahertz in power applications, but with decreasing power levels. Vacuum tube devices dominate high power (hundreds of kilowatts) at very high frequency (hundreds or thousands of megahertz) applications. Faster switching devices minimize energy lost in the transitions from on to off and back, but may create problems with radiated electromagnetic interference. Gate drive (or equivalent) circuits must be designed to supply sufficient drive current to achieve the full switching speed possible with a device. A device without sufficient drive to switch rapidly may be destroyed by excess heating.

Practical devices have non-zero voltage drop and dissipate power when on, and take some time to pass through an active region until they reach the "on" or "off" state. These losses are a significant part of the total lost power in a converter.

Power handling and dissipation of devices is also a critical factor in design. Power electronic devices may have to dissipate tens or hundreds of watts of waste heat, even switching as efficiently as possible between conducting and non-conducting states. In the switching mode, the power controlled is much larger than the power dissipated in the switch. The forward voltage drop in the conducting state translates into heat that must be dissipated. High power semiconductors require specialized heat sinks or active cooling systems to manage their junction temperature; exotic semiconductors such as silicon carbide have an advantage over straight silicon in this respect, and germanium, once the main-stay of solid-state electronics is now little used due to its unfavorable high temperature properties.

Semiconductor devices exist with ratings up to a few kilovolts in a single device. Where very high voltage must be controlled, multiple devices must be used in series, with networks to equalize voltage across all devices. Again, switching speed is a critical factor since the slowest-switching device will have to withstand a disproportionate share of the overall voltage. Mercury valves were once available with ratings to 100 kV in a single unit, simplifying their application in HVDC systems.

The current rating of a semiconductor device is limited by the heat generated within the dies and the heat developed in the resistance of the interconnecting leads. Semiconductor devices must be designed so that current is evenly distributed within the device across its internal junctions (or channels); once a "hot spot" develops, breakdown effects can rapidly destroy the device. Certain SCRs are available with current ratings to 3000 amperes in a single unit.

Solid-state Devices

DC/AC Converters (Inverters)

DC to AC converters produce an AC output waveform from a DC source. Applications include adjustable speed drives (ASD), uninterruptable power supplies (UPS), active filters, Flexible AC transmission systems (FACTS), voltage compensators, and photovoltaic generators. Topologies for these converters can be separated into two distinct categories: voltage source inverters and current source inverters. Voltage source inverters (VSIs) are named so because the independently controlled output is a voltage waveform. Similarly, current source inverters (CSIs) are distinct in that the controlled AC output is a current waveform.

Being static power converters, the DC to AC power conversion is the result of power switching devices, which are commonly fully controllable semiconductor power switches. The output waveforms are therefore made up of discrete values, producing fast transitions rather than smooth ones. The ability to produce near sinusoidal waveforms around the fundamental frequency is dictated by the modulation technique controlling when, and for how long, the power valves are on and off. Common modulation techniques include the carrier-based technique, or Pulse-width modulation, space-vector technique, and the selective-harmonic technique.

Voltage source inverters have practical uses in both single-phase and three-phase applications. Single-phase VSIs utilize half-bridge and full-bridge configurations, and are widely used for power supplies, single-phase UPSs, and elaborate high-power topologies when used in multicell configurations. Three-phase VSIs are used in applications that require sinusoidal voltage waveforms, such as ASDs, UPSs, and some types of FACTS devices such as the STATCOM. They are also used in applications where arbitrary voltages are required as in the case of active filters and voltage compensators.

Current source inverters are used to produce an AC output current from a DC current supply. This type of inverter is practical for three-phase applications in which high-quality voltage waveforms are required.

A relatively new class of inverters, called multilevel inverters, has gained widespread interest. Normal operation of CSIs and VSIs can be classified as two-level inverters, due to the fact that power switches connect to either the positive or to the negative DC bus. If more than two voltage levels were available to the inverter output terminals, the AC output could better approximate a sine wave. It is for this reason that multilevel inverters, although more complex and costly, offer higher performance.

Each inverter type differs in the DC links used, and in whether or not they require freewheeling diodes. Either can be made to operate in square-wave or pulse-width modulation (PWM) mode, depending on its intended usage. Square-wave mode offers simplicity, while PWM can be implemented several different ways and produces higher quality waveforms.

Voltage Source Inverters (VSI) feed the output inverter section from an approximately constant-voltage source.

The desired quality of the current output waveform determines which modulation technique needs to be selected for a given application. The output of a VSI is composed of discrete values. In order to obtain a smooth current waveform, the loads need to be inductive at the select harmonic frequencies. Without some sort of inductive filtering between the source and load, a capacitive load will cause the load to receive a choppy current waveform, with large and frequent current spikes.

There are three main types of VSIs:

1. Single-phase half-bridge inverter
2. Single-phase full-bridge inverter
3. Three-phase voltage source inverter

Single-phase Half-bridge Inverter

The single-phase voltage source half-bridge inverters, are meant for lower voltage applications and are commonly used in power supplies. Figure 9 shows the circuit schematic of this inverter.

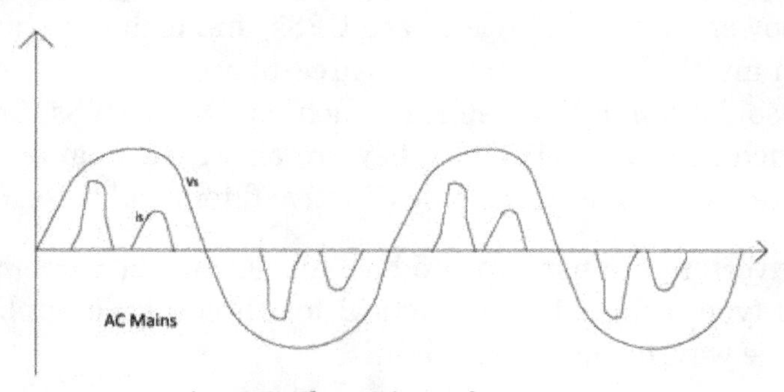

Figure 8: The AC input for an ASD.

Low-order current harmonics get injected back to the source voltage by the operation of the inverter. This means that two large capacitors are needed for filtering purposes in this design. As Figure 9 illustrates, only one switch can be on at time in each leg of the inverter. If both switches in a leg were on at the same time, the DC source will be shorted out.

Inverters can use several modulation techniques to control their switching schemes. The carrier-based PWM technique compares the AC output waveform, v_c, to a carrier voltage signal, v_Δ. When v_c is greater than v_Δ, S+ is on, and when v_c is less than v_Δ, S- is on. When the AC output is at frequency fc with its amplitude at v_c, and the triangular carrier signal is at frequency f_Δ with its amplitude at v_Δ, the PWM becomes a special sinusoidal case of the carrier based PWM. This case is dubbed sinusoidal pulse-width modulation (SPWM).For this, the modulation index, or amplitude-modulation ratio, is defined as $m_a = v_c/v_\Delta$.

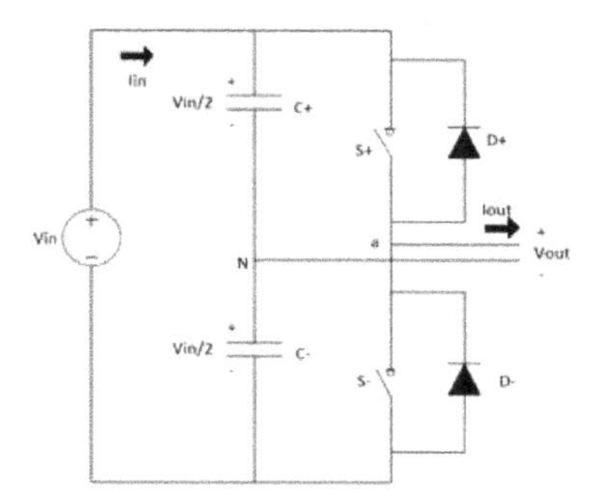

FIGURE 9: Single-Phase Half-Bridge Voltage Source Inverter

The normalized carrier frequency, or frequency-modulation ratio, is calculated using the equation $m_f = f_\Delta/f_c$.

If the over-modulation region, ma, exceeds one, a higher fundamental AC output voltage will be observed, but at the cost of saturation. For SPWM, the harmonics of the output waveform are at well-defined frequencies and amplitudes. This simplifies the design of the filtering components needed for the low-order current harmonic injection from the operation of the inverter. The maximum output amplitude in this mode of operation is half of the source voltage. If the maximum output amplitude, m_a, exceeds 3.24, the output waveform of the inverter becomes a square wave.

As was true for PWM, both switches in a leg for square wave modulation cannot be turned on at the same time, as this would cause a short across the voltage source. The switching scheme requires that both S+ and S- be on for a half cycle of the AC output period. The fundamental AC output amplitude is equal to $v_{o1} = v_{aN} = 2v_i/\pi$.

Its harmonics have an amplitude of $v_{oh} = v_{o1}/h$.

Therefore, the AC output voltage is not controlled by the inverter, but rather by the magnitude of the DC input voltage of the inverter.

Using selective harmonic elimination (SHE) as a modulation technique allows the switching of the inverter to selectively eliminate intrinsic harmonics. The fundamental component of the AC output voltage can also be adjusted within a desirable range. Since the AC output voltage obtained from this modulation technique has odd half and odd quarter wave symmetry, even harmonics do not exist. Any undesirable odd (N-1) intrinsic harmonics from the output waveform can be eliminated.

Single-phase Full-bridge Inverter

The full-bridge inverter is similar to the half bridge-inverter, but it has an additional leg to connect the neutral point to the load. Figure 3 shows the circuit schematic of the single-phase voltage source full-bridge inverter.

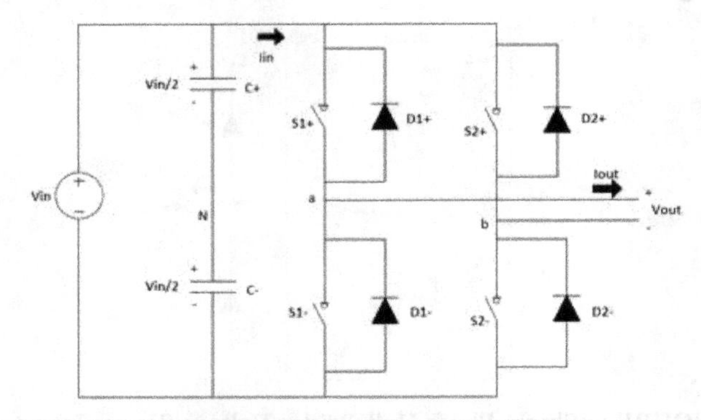

FIGURE 3: Single-Phase Voltage Source Full-Bridge Inverter

To avoid shorting out the voltage source, S1+ and S1- cannot be on at the same time, and S2+ and S2- also cannot be on at the same time. Any modulating technique used for the full-bridge configuration should have either the top or the bottom switch of each leg on at any given time. Due to the extra leg, the maximum amplitude of the output waveform is Vi, and is twice as large as the maximum achievable output amplitude for the half-bridge configuration.

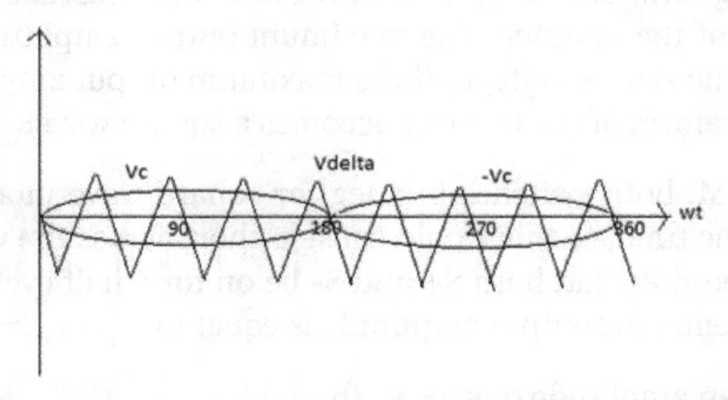

FIGURE 4: Carrier and Modulating Signals for the Bipolar Pulsewidth Modulation Technique

States 1 and 2 from Table 2 are used to generate the AC output voltage with bipolar SPWM. The AC output voltage can take on only two values, either Vi or –Vi. To generate these same states using a half-bridge configuration, a carrier based technique can be used. S+ being on for the half-bridge corresponds to S1+ and S2- being on for the full-bridge. Similarly, S- being on for the half-bridge corresponds to S1- and S2+ being on for the full bridge. The output voltage for this modulation technique is more or less sinusoidal, with a fundamental component that has an amplitude in the linear region of less than or equal to one $v_{01} = v_{ab1} = v_i \bullet m_a$.

Unlike the bipolar PWM technique, the unipolar approach uses states 1, 2, 3 and 4 from Table 2 to generate its AC output voltage. Therefore, the AC output voltage can take on the values Vi, 0 or –V i. To generate these states, two sinusoidal modulating signals, Vc and –Vc, are needed, as seen in Figure 4.

Vc is used to generate VaN, while –Vc is used to generate VbN. The following relationship is called unipolar carrier-based SPWM $v_{01} = 2 \bullet v_{aN1} = v_i \bullet m_a$.

The phase voltages VaN and VbN are identical, but 180 degrees out of phase with each other. The output voltage is equal to the difference of the two phase voltages, and do not contain any even harmonics. Therefore, if mf is taken, even the AC output voltage harmonics will appear at normalized odd frequencies, fh. These frequencies are centered on double the value of the normalized carrier frequency. This particular feature allows for smaller filtering components when trying to obtain a higher quality output waveform.

As was the case for the half-bridge SHE, the AC output voltage contains no even harmonics due to its odd half and odd quarter wave symmetry.

Three-phase Voltage Source Inverter

Single-phase VSIs are used primarily for low power range applications, while three-phase VSIs cover both medium and high power range applications. Figure 5 shows the circuit schematic for a three-phase VSI.

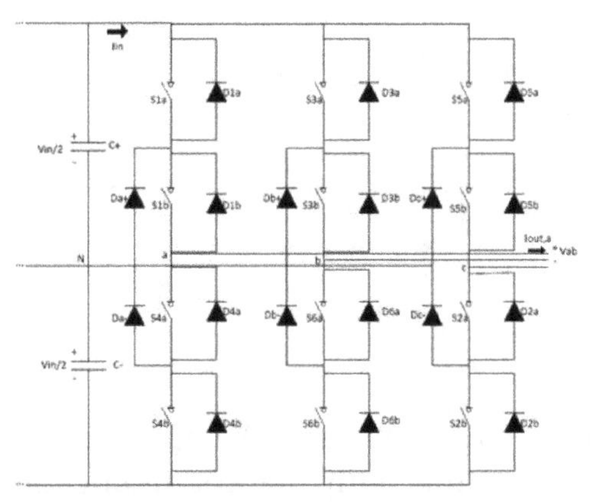

FIGURE 5: Three-Phase Voltage Source Inverter Circuit Schematic

FIGURE 6: Three-Phase Square-Wave Operation a) Switch State S1 b) Switch State S3 c) S1 Output d) S3 Output

Switches in any of the three legs of the inverter cannot be switched off simultaneously due to this resulting in the voltages being dependent on the respective line current's polarity. States 7 and 8 produce zero AC line voltages, which result in AC line currents freewheeling through either the upper or the lower components. However, the line voltages for states 1 through 6 produce an AC line voltage consisting of the discrete values of Vi, 0 or –Vi.

For three-phase SPWM, three modulating signals that are 120 degrees out of phase with one another are used in order to produce out of phase load voltages. In order to preserve the PWM features with a single carrier signal, the normalized carrier frequency, mf, needs to be a multiple of three. This keeps the magnitude of the phase voltages identical, but out of phase with each other by 120 degrees. The maximum achievable phase voltage amplitude in the linear region, ma less than or equal to one, is $v_{phase} = v_i / 2$. The maximum achievable line voltage amplitude is $V_{ab1} = v_{ab} \bullet \sqrt{3} / 2$

The only way to control the load voltage is by changing the input DC voltage.

Current Source Inverters

Current source inverters convert DC current into an AC current waveform. In applications requiring sinusoidal AC waveforms, magnitude, frequency, and phase should all be controlled. CSIs have high changes in current overtime, so capacitors are commonly employed on the AC side, while inductors are commonly employed on the DC side. Due to the absence of freewheeling diodes, the power circuit is reduced in size and weight, and tends to be more reliable than VSIs. Although single-phase topologies are possible, three-phase CSIs are more practical.

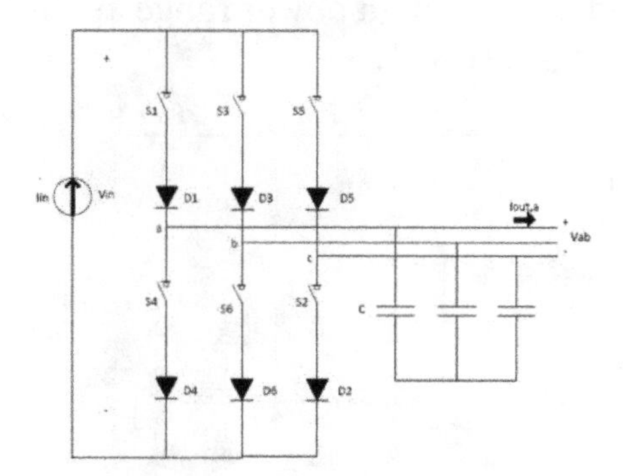

FIGURE 7: Three-Phase Current Source Inverter

In its most generalized form, a three-phase CSI employs the same conduction sequence as a six-pulse rectifier. At any time, only one common-cathode switch and one common-anode switch are on.

As a result, line currents take discrete values of –ii, 0 and ii. States are chosen such that a desired waveform is outputted and only valid states are used. This selection is based on modulating techniques, which include carrier-based PWM, selective harmonic elimination, and space-vector techniques.

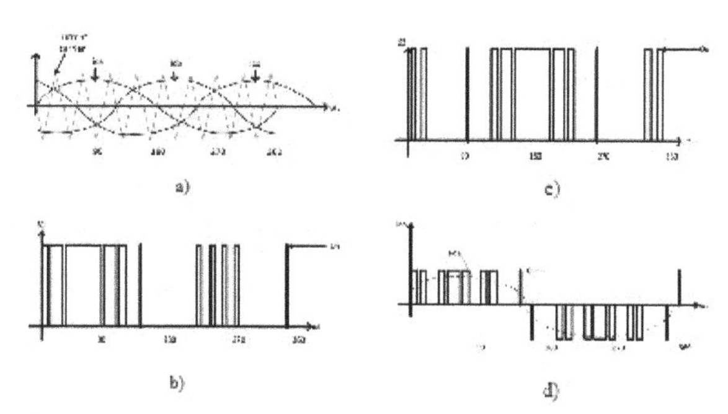

Figure 8: Synchronized-Pulse-Width-Modulation Waveforms for a Three-Phase Current Source Inverter a) Carrier and Modulating Signals b) S1 State c) S3 State d) Output Current

Carrier-based techniques used for VSIs can also be implemented for CSIs, resulting in CSI line currents that behave in the same way as VSI line voltages. The digital circuit utilized for modulating signals contains a switching pulse generator, a shorting pulse generator, a shorting pulse distributor, and a switching and shorting pulse combiner. A gating signal is produced based on a carrier current and three modulating signals.

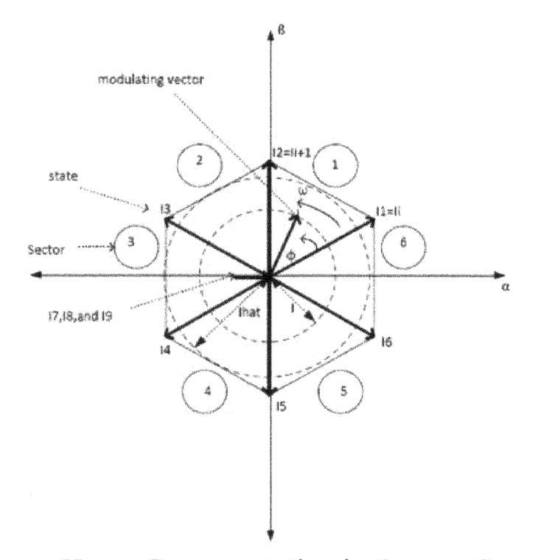

Figure 9: Space-Vector Representation in Current Source Inverters

A shorting pulse is added to this signal when no top switches and no bottom switches are gated, causing the RMS currents to be equal in all legs. The same methods are utilized for each phase, however, switching variables are 120 degrees out of phase relative to one another, and the current pulses are shifted by a half-cycle with respect to output currents. If a triangular carrier is used with sinusoidal modulating signals, the CSI is said to be utilizing synchronized-pulse-width-modulation (SPWM). If full over-modulation is used in conjunction with SPWM the inverter is said to be in square-wave operation.

The second CSI modulation category, SHE is also similar to its VSI counterpart. Utilizing the gating signals developed for a VSI and a set of synchronizing sinusoidal current signals, results in symmetrically distributed shorting pulses and, therefore, symmetrical gating patterns. This allows any arbitrary number of harmonics to be eliminated. It also allows control of the fundamental line current through the proper selection of primary switching angles. Optimal switching patterns must have quarter-wave and half-wave symmetry, as well as symmetry about 30 degrees and 150 degrees. Switching patterns are never allowed between 60 degrees and 120 degrees. The current ripple can be further reduced with the use of larger output capacitors, or by increasing the number of switching pulses.

The third category, space-vector-based modulation, generates PWM load line currents that equal load line currents, on average. Valid switching states and time selections are made digitally based on space vector transformation. Modulating signals are represented as a complex vector using a transformation equation. For balanced three-phase sinusoidal signals, this vector becomes a fixed module, which rotates at a frequency, ω. These space vectors are then used to approximate the modulating signal. If the signal is between arbitrary vectors, the vectors are combined with the zero vectors I7, I8, or I9. The following equations are used to ensure that the generated currents and the current vectors are on average equivalent.

Multilevel Inverters

A relatively new class called multilevel inverters has gained widespread interest. Normal operation of CSIs and VSIs can be classified as two-level inverters because the power switches connect to either the positive or the negative DC bus. If more than two voltage levels were available to the inverter output terminals, the AC output could better approximate a sine wave. For this reason multilevel inverters, although more complex and costly, offer higher performance. A three-level neutral-clamped inverter is shown in Figure 10.

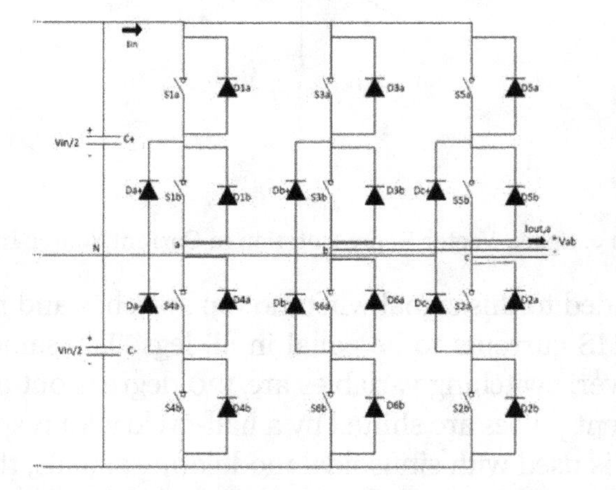

FIGURE 10: Three-Level Neutral-Clamped Inverter

Control methods for a three-level inverter only allow two switches of the four switches in each leg to simultaneously change conduction states. This allows smooth commutation and avoids shoot through by only selecting valid states. It may also be noted that since the DC bus voltage is shared by at least two power valves, their voltage ratings can be less than a two-level counterpart.

Carrier-based and space-vector modulation techniques are used for multilevel topologies. The methods for these techniques follow those of classic inverters, but with added complexity. Space-vector modulation offers a greater number of fixed voltage vectors to be used in approximating the modulation signal, and therefore allows more effective space vector PWM strategies to be accomplished at the cost of more elaborate algorithms. Due to added complexity and number of semiconductor devices, multilevel inverters are currently more suitable for high-power high-voltage applications. This technology reduces the harmonics hence improves overall efficiency of the scheme.

AC/AC Converters

Converting AC power to AC power allows control of the voltage, frequency, and phase of the waveform applied to a load from a supplied AC system . The two main categories that can be used to separate the types of converters are whether the frequency of the waveform is changed. AC/AC converter that don't allow the user to modify the frequencies are known as AC Voltage Controllers, or AC Regulators. AC converters that allow the user to change the frequency are simply referred to as frequency converters for AC to AC conversion. Under frequency converters there are three different types of converters that are typically used: cycloconverter, matrix converter, DC link converter (aka AC/DC/AC converter).

AC voltage controller: The purpose of an AC Voltage Controller, or AC Regulator, is to vary the RMS voltage across the load while at a constant frequency. Three control methods that are generally accepted are ON/OFF Control, Phase-Angle Control, and Pulse Width Modulation AC Chopper Control (PWM AC Chopper Control). All three of these methods can be implemented not only in single-phase circuits, but three-phase circuits as well.

- ON/OFF Control: Typically used for heating loads or speed control of motors, this control method involves turning the switch on for n integral cycles and turning the switch off for m integral cycles. Because turning the switches on and off causes undesirable harmonics to be created, the switches are turned on and off during zero-voltage and zero-current conditions (zero-crossing), effectively reducing the distortion.

- Phase-Angle Control: Various circuits exist to implement a phase-angle control on different waveforms, such as half-wave or full-wave voltage control. The power electronic components that are typically used are diodes, SCRs, and Triacs. With the use of these components, the user can delay the firing angle in a wave which will only cause part of the wave to be outputted.

- PWM AC Chopper Control: The other two control methods often have poor harmonics, output current quality, and input power factor. In order to improve these values PWM can be used instead of the other methods. What PWM AC Chopper does is have switches that turn on and off several times within alternate half-cycles of input voltage.

Matrix converters and cycloconverters: Cycloconverters are widely used in industry for ac to ac conversion, because they are able to be used in high-power applications. They are commutated direct frequency converters that are synchronised by a supply line. The cycloconverters output voltage waveforms have complex harmonics with the higher order harmonics being filtered by the machine inductance. Causing the machine current to have fewer harmonics, while the remaining harmonics causes losses and torque pulsations. Note that in a cycloconverter, unlike other converters, there are no inductors or capacitors, i.e. no storage devices. For this reason, the instantaneous input power and the output power are equal.

- Single-Phase to Single-Phase Cycloconverters: Single-Phase to Single-Phase Cycloconverters started drawing more interest recently because of the decrease in both size and price of the power electronics switches. The single-phase high frequency ac voltage can be either sinusoidal or trapezoidal. These might be zero voltage intervals for control purpose or zero voltage commutation.
- Three-Phase to Single-Phase Cycloconverters: There are two kinds of three-phase to single-phase cycloconverters: 3φ to 1φ half wave cycloconverters and 3φ to 1φ bridge cycloconverters. Both positive and negative converters can generate voltage at either polarity, resulting in the positive converter only supplying positive current, and the negative converter only supplying negative current.

With recent device advances, newer forms of cycloconverters are being developed, such as matrix converters. The first change that is first noticed is that matrix converters utilize bi-directional, bipolar switches. A single phase to a single phase matrix converter consists of a matrix of 9 switches connecting the three input phases to the tree output phase. Any input phase and output phase can be connected together at any time without connecting any two switches from the same phase at the same time; otherwise this will cause a short circuit of the input phases. Matrix converters are lighter, more compact and versatile than other converter solutions. As a result, they are able to achieve higher levels of integration, higher temperature operation, broad output frequency and natural bi-directional power flow suitable to regenerate energy back to the utility.

The matrix converters are subdivided into two types: direct and indirect converters. A direct matrix converter with three-phase input and three-phase output, the switches in a matrix converter must be bi-directional, that is, they must be able to block voltages of either polarity and to conduct current in either direction. This switching strategy permits the highest possible output voltage and reduces the reactive line-side current. Therefore, the power flow through the converter is reversible. Because of its commutation problem and complex control keep it from being broadly utilized in industry.

Unlike the direct matrix converters, the indirect matrix converters has the same functionality, but uses separate input and output sections that are connected through a dc link without storage elements. The design includes a four-quadrant current source rectifier and a voltage source inverter. The input section consists of bi-directional bipolar switches. The commutation strategy can be applied by changing the switching state of the input section while the output section is in a freewheeling mode. This commutation algorithm is significantly less complexity and higher reliability as compared to a conventional direct matrix converter.

DC link converters: DC Link Converters, also referred to as AC/DC/AC converters, convert an AC input to an AC output with the use of a DC link in the middle. Meaning that the power in the converter is converted to DC from AC with the use of a rectifier, and then it is converted back to AC from DC with the use of an inverter. The end result is an output with a lower voltage and variable (higher or lower) frequency. Due to their wide area of application, the AC/DC/AC converters are the most common contemporary solution. Other advantages to AC/DC/AC converters is that they are stable in overload and no-load conditions, as well as they can be disengaged from a load without damage.

Hybrid matrix converter: Hybrid matrix converters are relatively new for AC/AC converters. These converters combine the AC/DC/AC design with the matrix converter design. Multiple types of hybrid converters have been developed in this new category, an example being a converter that uses uni-directional switches and two converter stages without the dc-link; without the capacitors or inductors needed for a dc-link, the weight and size of the converter is reduced. Two sub-categories exist from the hybrid converters, named hybrid direct matrix converter (HDMC) and hybrid indirect matrix converter (HIMC). HDMC convert the voltage and current in one stage, while the HIMC utilizes separate stages, like the AC/DC/AC converter, but without the use of an intermediate storage element.

Applications: Below is a list of common applications that each converter is used in.

- AC Voltage Controller: Lighting Control; Domestic and Industrial Heating; Speed Control of Fan,Pump or Hoist Drives, Soft Starting of Induction Motors, Static AC Switches (Temperature Control, Transformer Tap Changing, etc.)
- Cycloconverter: High-Power Low-Speed Reversible AC Motor Drives; Constant Frequency Power Supply with Variable Input Frequency; Controllable VAR Generators for Power Factor Correction; AC System Interties Linking Two Independent Power Systems.
- Matrix Converter: Currently the application of matrix converters are limited due to non-availability of bilateral monolithic switches capable of operating at high frequency, complex control law implementation, commutation and other reasons. With these developments, matrix converters could replace cycloconverters in many areas.
- DC Link: Can be used for individual or multiple load applications of machine building and construction.

Simulations of Power Electronic Systems

Power electronic circuits are simulated using computer simulation programs such as PLECS, PSIM and MATLAB/simulink. Circuits are simulated before they are produced to test how the circuits respond under certain conditions. Also, creating a simulation is both cheaper and faster than creating a prototype to use for testing.

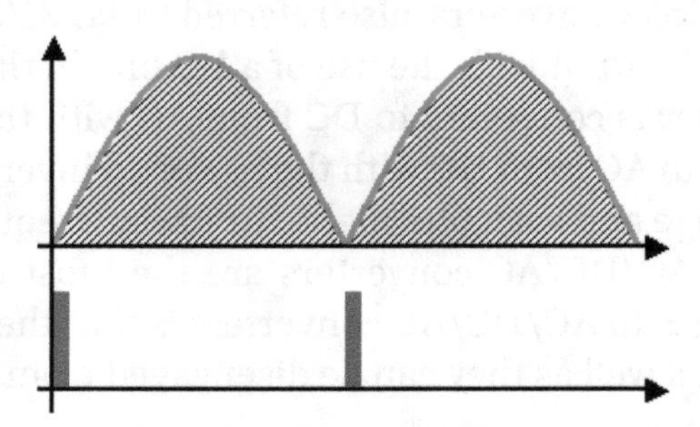

Output voltage of a full-wave rectifier with controlled thyristors

Applications

Applications of power electronics range in size from a switched mode power supply in an AC adapter, battery chargers, audio amplifiers, fluorescent lamp ballasts, through variable frequency drives and DC motor drives used to operate pumps, fans, and manufacturing machinery, up to gigawatt-scale high voltage direct current power transmission systems used to interconnect electrical grids. Power electronic systems are found in virtually every electronic device. For example:

- DC/DC converters are used in most mobile devices (mobile phones, PDA etc.) to maintain the voltage at a fixed value whatever the voltage level of the battery is. These converters are also used for electronic isolation and power factor correction. A power optimizer is a type of DC/DC converter developed to maximize the energy harvest from solar photovoltaic or wind turbine systems.

- AC/DC converters (rectifiers) are used every time an electronic device is connected to the mains (computer, television etc.). These may simply change AC to DC or can also change the voltage level as part of their operation.

- AC/AC converters are used to change either the voltage level or the frequency (international power adapters, light dimmer). In power distribution networks AC/AC converters may be used to exchange power between utility frequency 50 Hz and 60 Hz power grids.

- DC/AC converters (inverters) are used primarily in UPS or renewable energy systems or emergency lighting systems. Mains power charges the DC battery. If the mains fails, an inverter produces AC electricity at mains voltage from the DC battery. Solar inverter, both smaller string and larger central inverters, as well as solar micro-inverter are used in photovoltaics as a component of a PV system.

Motor drives are found in pumps, blowers, and mill drives for textile, paper, cement and other such facilities. Drives may be used for power conversion and for motion control. For AC motors, applications include variable-frequency drives, motor soft starters and excitation systems.

In hybrid electric vehicles (HEVs), power electronics are used in two formats: series hybrid and parallel hybrid. The difference between a series hybrid and a parallel hybrid is the relationship of the electric motor to the internal combustion engine (ICE). Devices used in electric vehicles consist mostly of dc/dc converters for battery charging and dc/ac converters to power the propulsion motor. Electric trains use power electronic devices to obtain power, as well as for vector control using pulse width modulation (PWM) rectifiers. The trains obtain their power from power lines. Another new usage for power electronics is in elevator systems. These systems may use thyristors, inverters, permanent magnet motors, or various hybrid systems that incorporate PWM systems and standard motors.

Inverters

In general, inverters are utilized in applications requiring direct conversion of electrical energy from DC to AC or indirect conversion from AC to AC. DC to AC conversion is useful for many fields, including power conditioning, harmonic compensation, motor drives, and renewable energy grid-integration.

In power systems it is often desired to eliminate harmonic content found in line currents. VSIs can be used as active power filters to provide this compensation. Based on measured line currents and voltages, a control system determines reference current signals for each phase. This is fed back through an outer loop and subtracted from actual current signals to create current signals for an inner loop to the inverter. These signals then cause the inverter to generate output currents that compensate for the harmonic content. This configuration requires no real power consumption, as it is fully fed by the line; the DC link is simply a capacitor that is kept at a constant voltage by the control system. In this configuration, output currents are in phase with line voltages to produce a unity power factor. Conversely, VAR compensation is possible in a similar configuration where output currents lead line voltages to improve the overall power factor.

In facilities that require energy at all times, such as hospitals and airports, UPS systems are utilized. In a standby system, an inverter is brought online when the normally supplying grid is interrupted. Power is instantaneously drawn from onsite batteries and converted into usable AC voltage by the VSI, until grid power is restored, or until backup generators are brought online. In an online UPS system, a rectifier-DC-link-inverter is used to protect the load from transients and harmonic content. A battery in parallel with the DC-link is kept fully charged by the output in case the grid power is interrupted, while the output of the inverter is fed through a low pass filter to the load. High power quality and independence from disturbances is achieved.

Various AC motor drives have been developed for speed, torque, and position control of AC motors. These drives can be categorized as low-performance or as high-performance, based on whether they are scalar-controlled or vector-controlled, respectively. In scalar-controlled drives, fundamental stator current, or voltage frequency and amplitude, are the only controllable quantities. Therefore, these drives are employed in applications where high quality control is not required, such as fans and compressors. On the other hand, vector-controlled drives allow for instantaneous current and voltage values to be controlled continuously. This high performance is necessary for applications such as elevators and electric cars.

Inverters are also vital to many renewable energy applications. In photovoltaic purposes, the inverter, which is usually a PWM VSI, gets fed by the DC electrical energy output of a photovoltaic module or array. The inverter then converts this into an AC voltage to be interfaced with either a load or the utility grid. Inverters may also be employed in other renewable systems, such as wind turbines. In these applications, the turbine speed usually varies causing changes in voltage frequency and sometimes in the magnitude. In this case, the generated voltage can be rectified and then inverted to stabilize frequency and magnitude.

Smart Grid

A smart grid is a modernized electrical grid that uses information and communications technology to gather and act on information, such as information about the behaviors of suppliers and consumers, in an automated fashion to improve the efficiency, reliability, economics, and sustainability of the production and distribution of electricity.

Electric power generated by wind turbines and hydroelectric turbines by using induction generators can cause variances in the frequency at which power is generated. Power electronic devices are utilized in these systems to convert the generated ac voltages into high-voltage direct current (HVDC). The HVDC power can be more easily converted into three phase power that is coherent with the power associated to the existing power grid. Through these devices, the power delivered by these systems is cleaner and has a higher associated power factor. Wind power systems optimum torque is obtained either through a gearbox or direct drive technologies that can reduce the size of the power electronics device.

Electric power can be generated through photovoltaic cells by using power electronic devices. The produced power is usually then transformed by solar inverters. Inverters are divided into three different types: central, module-integrated and string. Central converters can be connected either in parallel or in series on the DC side of the system. For photovoltaic "farms", a single central converter is used for the entire system. Module-integrated converters are connected in series on either the DC or AC side. Normally several modules are used within a photovoltaic system, since the system requires these converters on both DC and AC terminals. A string converter is used in a system that utilizes photovoltaic cells that are facing different directions. It is used to convert the power generated to each string, or line, in which the photovoltaic cells are interacting.

Grid Voltage Regulation

Power electronics can be used to help utilities adapt to the rapid increase in distributed residential/commercial solar power generation. Germany and parts of Hawaii, California and New Jersey require costly studies to be conducted before approving new solar installations. Relatively small-scale ground- or pole-mounted devices create the potential for a distributed control infrastructure to monitor and manage the flow of power. Traditional electromechanical systems, such as capacitor banks or voltage regulators at substations, can take minutes to adjust voltage and can be distant from the solar installations where the problems originate. If voltage on a neighborhood circuit goes too high, it can endanger utility crews and cause damage to both utility and customer equipment. Further, a grid fault causes photovoltaic generators to shut down immediately, spiking demand for grid power. Smart grid-based regulators are more controllable than far more numerous consumer devices.

In another approach, a group of 16 western utilities called the Western Electric Industry Leaders called for mandatory use of "smart inverters". These devices convert DC to household AC and can also help with power quality. Such devices could eliminate the need for expensive utility equipment upgrades at a much lower total cost.

DC-to-DC Converter

A DC-to-DC converter is an electronic circuit or electromechanical device that converts a source of direct current (DC) from one voltage level to another. It is a type of electric power converter. Power levels range from very low (small batteries) to very high (high-voltage power transmission).

History

Before the development of power semiconductors and allied technologies, one way to convert the voltage of a DC supply to a higher voltage, for low-power applications, was to convert it to AC by using a vibrator, followed by a step-up transformer and rectifier. For higher power an electric motor was used to drive a generator of the desired voltage (sometimes combined into a single "dynamotor" unit, a motor and generator combined into one unit, with one winding driving the motor and the other generating the output voltage). These were relatively inefficient and expensive procedures used only when there was no alternative, as to power a car radio (which then used thermionic valves/ tubes requiring much higher voltages than available from a 6 or 12 V car battery). The introduction of power semiconductors and integrated circuits made it economically viable to use techniques as described below, for example to convert the DC power supply to high-frequency AC, use a transformer—small, light, and cheap due to the high frequency—to change the voltage, and rectify back to DC. Although by 1976 transistor car radio receivers did not require high voltages, some amateur radio operators continued to use vibrator supplies and dynamotors for mobile transceivers requiring high voltages, although transistorised power supplies were available.

While it was possible to derive a lower voltage from a higher with a linear electronic circuit, or even a resistor, these methods dissipated the excess as heat; energy-efficient conversion only became possible with solid-state switch-mode circuits.

Uses

DC to DC converters are used in portable electronic devices such as cellular phones and laptop computers, which are supplied with power from batteries primarily. Such electronic devices often contain several sub-circuits, each with its own voltage level requirement different from that supplied by the battery or an external supply (sometimes higher or lower than the supply voltage). Additionally, the battery voltage declines as its stored energy is drained. Switched DC to DC converters offer a method to increase voltage from a partially lowered battery voltage thereby saving space instead of using multiple batteries to accomplish the same thing.

Most DC to DC converter circuits also regulate the output voltage. Some exceptions include high-efficiency LED power sources, which are a kind of DC to DC converter that regulates the current through the LEDs, and simple charge pumps which double or triple the output voltage.

DC to DC converters developed to maximize the energy harvest for photovoltaic systems and for wind turbines are called power optimizers.

Transformers used for voltage conversion at mains frequencies of 50–60 Hz must be large and heavy for powers exceeding a few watts. This makes them expensive, and they are subject to energy losses in their windings and due to eddy currents in their cores. DC-to-DC techniques that use transformers or inductors work at much higher frequencies, requiring only much smaller, lighter, and cheaper wound components. Consequently these techniques are used even where a mains transformer could be used; for example, for domestic electronic appliances it is preferable to rectify mains voltage to DC, use switch-mode techniques to convert it to high-frequency AC at the desired voltage, then, usually, rectify to DC. The entire complex circuit is cheaper and more efficient than a simple mains transformer circuit of the same output.

Electronic Conversion

Linear regulators which are used to output a stable DC independent of input voltage and output load from a higher but less stable input by dissipating excess volt-amperes as heat, could be described literally as DC-to-DC converters, but this is not usual usage. (The same could be said of a simple voltage dropper resistor, whether or not stabilised by a following voltage regulator or Zener diode.)

There are also simple capacitive voltage doubler and Dickson multiplier circuits using diodes and capacitors to multiply a DC voltage by an integer value, typically delivering only a small current.

Practical electronic converters use switching techniques. Switched-mode DC-to-DC converters convert one DC voltage level to another, which may be higher or lower, by storing the input energy temporarily and then releasing that energy to the output at a different voltage. The storage may be in either magnetic field storage components (inductors, transformers) or electric field storage components (capacitors). This conversion method can increase or decrease voltage. Switching conversion is more power efficient (often 75% to 98%) than linear voltage regulation, which dissipates unwanted power as heat. Fast semiconductor device rise and fall times are required for efficiency; however, these fast transitions combine with layout parasitic effects to make circuit design challenging. The higher efficiency of a switched-mode converter reduces the heatsinking needed, and increases battery endurance of portable equipment. Efficiency has improved since the late 1980s due to the use of power FETs, which are able to switch more efficiently with lower switching losses at higher frequencies than power bipolar transistors, and use less complex drive circuitry. Another important improvement in DC-DC converters is replacing the flywheel diode by synchronous rectification using a power FET, whose "on resistance" is much lower, reducing switching losses. Before the wide availability of power semiconductors, low-power DC-to-DC synchronous converters consisted of an electro-mechanical vibrator followed by a voltage step-up transformer feeding a vacuum tube or semiconductor rectifier, or synchronous rectifier contacts on the vibrator.

Most DC-to-DC converters are designed to move power in only one direction, from dedicated input to output. However, all switching regulator topologies can be made bidirectional and able to move power in either direction by replacing all diodes with independently controlled active rectification. A bidirectional converter is useful, for example, in applications requiring regenerative braking of vehicles, where power is supplied to the wheels while driving, but supplied by the wheels when braking.

Switching converters are electronically complex, although this is embodied in integrated circuits, with few components needed. They need careful design of the circuit and physical layout to reduce switching noise (EMI / RFI) to acceptable levels and, like all high-frequency circuits, for stable operation. Cost was higher than linear regulators in voltage-dropping applications, but this dropped with advances in chip design.

DC-to-DC converters are available as integrated circuits (ICs) requiring few additional components. Converters are also available as complete hybrid circuit modules, ready for use within an electronic assembly.

Magnetic

In these DC-to-DC converters, energy is periodically stored within and released from a magnetic field in an inductor or a transformer, typically within a frequency range of 300 kHz to 10 MHz. By adjusting the duty cycle of the charging voltage (that is, the ratio of the on/off times), the amount of power transferred to a load can be more easily controlled, though this control can also be applied to the input current, the output current, or to maintain constant power. Transformer-based converters may provide isolation between input and output. In general, the term DC-to-DC converter refers to one of these switching converters. These circuits are the heart of a switched-mode power supply. Many topologies exist. This table shows the most common ones.

	Forward (energy transfers through the magnetic field)	**Flyback (energy is stored in the magnetic field)**
No transformer (non-isolated)	• Step-down (buck) - The output voltage is lower than the input voltage, and of the same polarity.	• Non-inverting: The output voltage is the same polarity as the input. ○ Step-up (boost) - The output voltage is higher than the input voltage. ○ SEPIC - The output voltage can be lower or higher than the input. • Inverting: the output voltage is of the opposite polarity as the input. ○ Inverting (buck-boost). ○ Ćuk - Output current is continuous.
	• True buck-boost - The output voltage is the same polarity as the input and can be lower or higher.	
	• Split-pi (boost-buck) - Allows bidirectional voltage conversion with the output voltage the same polarity as the input and can be lower or higher.	

With trans-former (isolatable)	• Forward - 1 or 2 transistor drive. • Push-pull (half bridge) - 2 transistors drive. • Full bridge - 4 transistor drive.	• Flyback - 1 transistor drive.

In addition, each topology may be:

Hard switched

Transistors switch quickly while exposed to both full voltage and full current

Resonant

An LC circuit shapes the voltage across the transistor and current through it so that the transistor switches when either the voltage or the current is zero

Magnetic DC-to-DC converters may be operated in two modes, according to the current in its main magnetic component (inductor or transformer):

Continuous

The current fluctuates but never goes down to zero

Discontinuous

The current fluctuates during the cycle, going down to zero at or before the end of each cycle

A converter may be designed to operate in continuous mode at high power, and in discontinuous mode at low power.

The half bridge and flyback topologies are similar in that energy stored in the magnetic core needs to be dissipated so that the core does not saturate. Power transmission in a flyback circuit is limited by the amount of energy that can be stored in the core, while forward circuits are usually limited by the I/V characteristics of the switches.

Although MOSFET switches can tolerate simultaneous full current and voltage (although thermal stress and electromigration can shorten the MTBF), bipolar switches generally can't so require the use of a snubber (or two).

High-current systems often use multiphase converters, also called interleaved converters. Multiphase regulators can have better ripple and better response times than single-phase regulators.

Many laptop and desktop motherboards include interleaved buck regulators, sometimes as a voltage regulator module.

Capacitive

Switched capacitor converters rely on alternately connecting capacitors to the input and output in differing topologies. For example, a switched-capacitor reducing converter might charge two capacitors in series and then discharge them in parallel. This would produce the same output power (less that lost to efficiency of under 100%) at, ideally, half the input voltage and twice the current. Because they operate on discrete quantities of charge, these are also sometimes referred to as charge pump converters. They are typically used in applications requiring relatively small currents, as at higher currents the increased efficiency and smaller size of switch-mode converters makes them a better choice. They are also used at extremely high voltages, as magnetics would break down at such voltages.

Electromechanical Conversion

A motor-generator set, mainly of historical interest, consists of an electric motor and generator coupled together. A dynamotor combines both functions into a single unit with coils for both the motor and the generator functions wound around a single rotor; both coils share the same outer field coils or magnets. Typically the motor coils are driven from a commutator on one end of the shaft, when the generator coils output to another commutator on the other end of the shaft. The entire rotor and shaft assembly is smaller in size than a pair of machines, and may not have any exposed drive shafts.

A motor generator with separate motor and generator.

Motor-generators can convert between any combination of DC and AC voltage and phase standards. Large motor-generator sets were widely used to convert industrial amounts of power while smaller units were used to convert battery power (6, 12 or 24 V DC) to a high DC voltage, which was required to operate vacuum tube (thermionic valve) equipment.

For lower-power requirements at voltages higher than supplied by a vehicle battery, vibrator or "buzzer" power supplies were used. The vibrator oscillated mechanically, with contacts that switched the polarity of the battery many times per second, effectively converting DC to square wave AC, which could then be fed to a transformer of the required output voltage(s). It made a characteristic buzzing noise.

Electrochemical Conversion

A further means of DC to DC conversion in the kilowatts to megawatts range is presented by using redox flow batteries such as the vanadium redox battery.

Chaotic Behavior

DC-to-DC converters are subject to different types of chaotic dynamics such as bifurcation, crisis, and intermittency.

Terminology

Step-down

A converter where output voltage is lower than the input voltage (such as a buck converter).

Step-up

A converter that outputs a voltage higher than the input voltage (such as a boost converter).

Continuous Current Mode

Current and thus the magnetic field in the inductive energy storage never reach zero.

Discontinuous Current Mode

Current and thus the magnetic field in the inductive energy storage may reach or cross zero.

Noise

Unwanted electrical and electromagnetic signal noise, typically switching artifacts.

RF Noise

Switching converters inherently emit radio waves at the switching frequency and its harmonics. Switching converters that produce triangular switching current, such as the Split-Pi, forward converter, or Ćuk converter in continuous current mode, produce less harmonic noise than other switching converters. RF noise causes electromagnetic interference (EMI). Acceptable levels depend upon requirements, e.g. proximity to RF circuitry needs more suppression than simply meeting regulations.

Input Noise

The input voltage may have non-negligible noise. Additionally, if the converter loads the input with sharp load edges, the converter can emit RF noise from the supplying power lines. This should be prevented with proper filtering in the input stage of the converter.

Output Noise

The output of an ideal DC-to-DC converter is a flat, constant output voltage. However, real converters produce a DC output upon which is superimposed some level of electrical noise. Switching converters produce switching noise at the switching frequency and its harmonics. Additionally, all electronic circuits have some thermal noise. Some sensitive radio-frequency and analog circuits require a power supply with so little noise that it can only be provided by a linear regulator. Some analog circuits which require a power supply with relatively low noise can tolerate some of the less-noisy switching converters, e.g. using continuous triangular waveforms rather than square waves.

Rectifier

A rectifier is an electrical device that converts alternating current (AC), which periodically reverses direction, to direct current (DC), which flows in only one direction. The process is known as rectification. Physically, rectifiers take a number of forms, including vacuum tube diodes, mercury-arc valves, copper and selenium oxide rectifiers, semiconductor diodes, silicon-controlled rectifiers and other silicon-based semiconductor switches. Historically, even synchronous electromechanical switches and motors have been used. Early radio receivers, called crystal radios, used a "cat's whisker" of fine wire pressing on a crystal of galena (lead sulfide) to serve as a point-contact rectifier or "crystal detector".

A rectifier diode (silicon controlled rectifier) and associated mounting hardware. The heavy threaded stud attaches the device to a heatsink to dissipate heat.

Rectifiers have many uses, but are often found serving as components of DC power supplies and high-voltage direct current power transmission systems. Rectification may serve in roles other than to generate direct current for use as a source of power. As noted, detectors of radio signals serve as rectifiers. In gas heating systems flame rectification is used to detect presence of a flame.

Because of the alternating nature of the input AC sine wave, the process of rectification alone produces a DC current that, though unidirectional, consists of pulses of current. Many applications of rectifiers, such as power supplies for radio, television and computer equipment, require a steady constant DC current (as would be produced by a battery). In these applications the output of the rectifier is smoothed by an electronic filter (usually a capacitor) to produce a steady current.

More complex circuitry that performs the opposite function, converting DC to AC, is called an inverter.

Rectifier Devices

Before the development of silicon semiconductor rectifiers, vacuum tube thermionic diodes and copper oxide- or selenium-based metal rectifier stacks were used. With the introduction of semiconductor electronics, vacuum tube rectifiers became obsolete, except for some enthusiasts of vacuum tube audio equipment. For power rectification from very low to very high current, semiconductor diodes of various types (junction diodes, Schottky diodes, etc.) are widely used.

Other devices that have control electrodes as well as acting as unidirectional current valves are used where more than simple rectification is required—e.g., where variable output voltage is needed. High-power rectifiers, such as those used in high-voltage direct current power transmission, employ silicon semiconductor devices of various types. These are thyristors or other controlled switching solid-state switches, which effectively function as diodes to pass current in only one direction.

Rectifier Circuits

Rectifier circuits may be single-phase or multi-phase (three being the most common number of phases). Most low power rectifiers for domestic equipment are single-phase, but three-phase rectification is very important for industrial applications and for the transmission of energy as DC (HVDC).

Single-phase Rectifiers

Half-wave Rectification

In half-wave rectification of a single-phase supply, either the positive or negative half of the AC wave is passed, while the other half is blocked. Because only one half of the input waveform reaches the output, mean voltage is lower. Half-wave rectification requires a single diode in a single-phase supply, or three in a three-phase supply. Rectifiers yield a unidirectional but pulsating direct current; half-wave rectifiers produce far more ripple than full-wave rectifiers, and much more filtering is needed to eliminate harmonics of the AC frequency from the output.

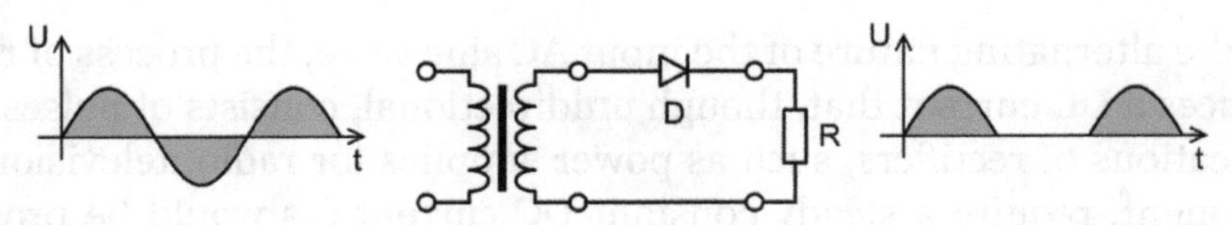

Half-wave Rectifier

The no-load output DC voltage of an ideal half-wave rectifier for a sinusoidal input voltage is:

$$dV_{\mathrm{rms}} = \frac{V_{\mathrm{peak}}}{2}[8pt]V_{\mathrm{dc}} = \frac{V_{\mathrm{peak}}}{\pi}$$

where:

V_{dc}, V_{av} – the DC or average output voltage,

V_{peak}, the peak value of the phase input voltages,

V_{rms}, the root mean square (RMS) value of output voltage.

Full-wave Rectification

A full-wave rectifier converts the whole of the input waveform to one of constant polarity (positive or negative) at its output. Full-wave rectification converts both polarities of the input waveform to pulsating DC (direct current), and yields a higher average output voltage. Two diodes and a center tapped transformer, or four diodes in a bridge configuration and any AC source (including a transformer without center tap), are needed. Single semiconductor diodes, double diodes with common cathode or common anode, and four-diode bridges, are manufactured as single components.

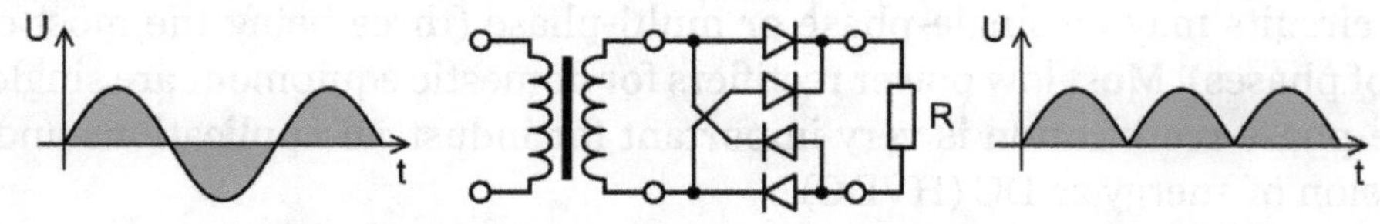

Graetz bridge rectifier: a full-wave rectifier using four diodes.

For single-phase AC, if the transformer is center-tapped, then two diodes back-to-back (cathode-to-cathode or anode-to-anode, depending upon output polarity required) can form a full-wave rectifier. Twice as many turns are required on the transformer secondary to obtain the same output voltage than for a bridge rectifier, but the power rating is unchanged.

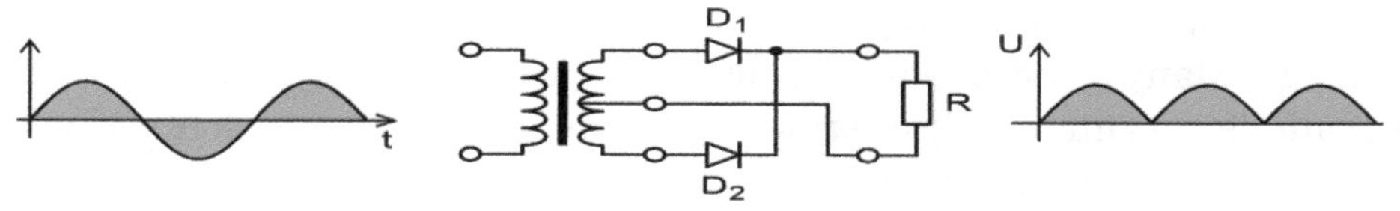

Full-wave rectifier using a center tap transformer and 2 diodes.

The average and RMS no-load output voltages of an ideal single-phase full-wave rectifier are:

$$V_{dc} = V_{av} = \frac{2V_{peak}}{\pi}[8pt]V_{rms} = \frac{V_{peak}}{\sqrt{2}}$$

Very common double-diode rectifier vacuum tubes contained a single common cathode and two anodes inside a single envelope, achieving full-wave rectification with positive output. The 5U4 and 5Y3 were popular examples of this configuration.

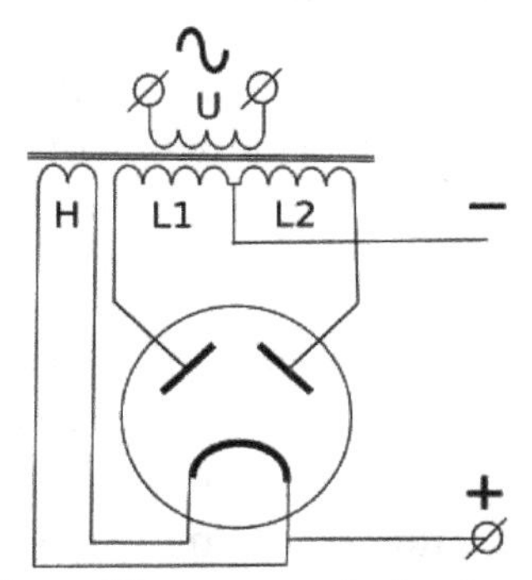

Full-wave rectifier, with vacuum tube having two anodes.

Three-phase Rectifiers

Single-phase rectifiers are commonly used for power supplies for domestic equipment. However, for most industrial and high-power applications, three-phase rectifier circuits are the norm. As with single-phase rectifiers, three-phase rectifiers can take the form of a half-wave circuit, a full-wave circuit using a center-tapped transformer, or a full-wave bridge circuit.

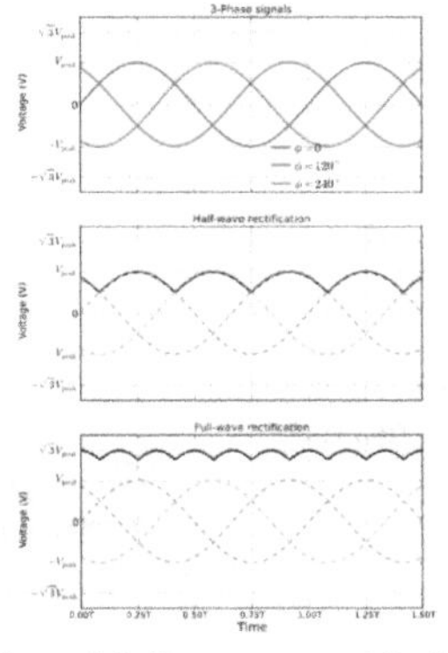

3-phase AC input, half- and full-wave rectified DC output waveforms

Thyristors are commonly used in place of diodes to create a circuit that can regulate the output voltage. Many devices that provide direct current actually generate three-phase AC. For example, an automobile alternator contains six diodes, which function as a full-wave rectifier for battery charging.

Three-phase, Half-wave Circuit

An uncontrolled three-phase, half-wave circuit requires three diodes, one connected to each phase. This is the simplest type of three-phase rectifier but suffers from relatively high harmonic distortion on both the AC and DC connections. This type of rectifier is said to have a pulse-number of three, since the output voltage on the DC side contains three distinct pulses per cycle of the grid frequency.

Three-phase, full-wave circuit using center-tapped transformer

If the AC supply is fed via a transformer with a center tap, a rectifier circuit with improved harmonic performance can be obtained. This rectifier now requires six diodes, one connected to each end of each transformer secondary winding. This circuit has a pulse-number of six, and in effect, can be thought of as a six-phase, half-wave circuit.

Before solid state devices became available, the half-wave circuit, and the full-wave circuit using a center-tapped transformer, were very commonly used in industrial rectifiers using mercury-arc valves. This was because the three or six AC supply inputs could be fed to a corresponding number of anode electrodes on a single tank, sharing a common cathode.

With the advent of diodes and thyristors, these circuits have become less popular and the three-phase bridge circuit has become the most common circuit.

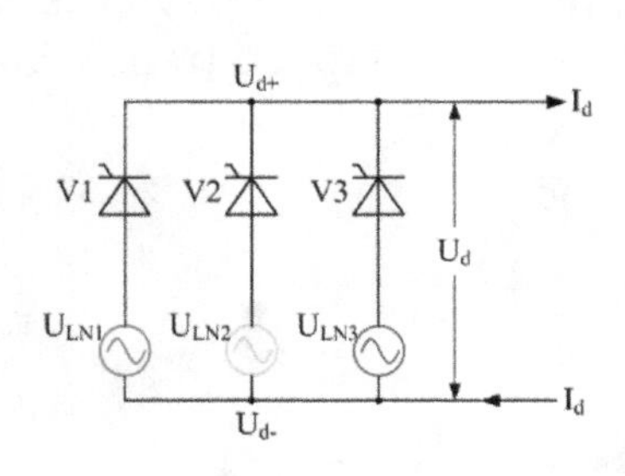

Three-phase half-wave rectifier circuit using thyristors as the switching elements, ignoring supply inductance

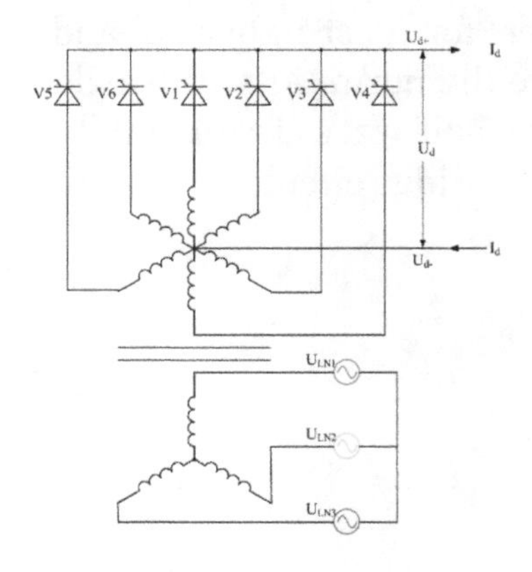

Three-phase full-wave rectifier circuit using thyristors as the switching elements, with a center-tapped transformer, ignoring supply inductance

Three-phase Bridge Rectifier

For an uncontrolled three-phase bridge rectifier, six diodes are used, and the circuit again has a pulse number of six. For this reason, it is also commonly referred to as a six-pulse bridge.

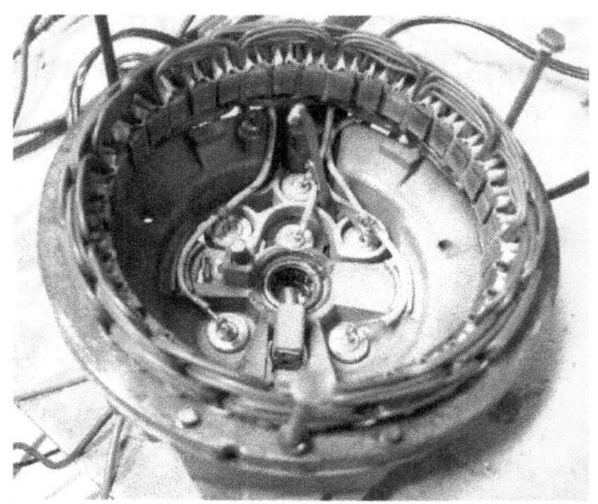

Disassembled automobile alternator, showing the six diodes that comprise a full-wave three-phase bridge rectifier.

For low-power applications, double diodes in series, with the anode of the first diode connected to the cathode of the second, are manufactured as a single component for this purpose. Some commercially available double diodes have all four terminals available so the user can configure them for single-phase split supply use, half a bridge, or three-phase rectifier.

For higher-power applications, a single discrete device is usually used for each of the six arms of the bridge. For the very highest powers, each arm of the bridge may consist of tens or hundreds of separate devices in parallel (where very high current is needed, for example in aluminium smelting) or in series (where very high voltages are needed, for example in high-voltage direct current power transmission).

For a three-phase full-wave diode rectifier, the ideal, no-load average output voltage is

$$V_{dc} = V_{av} = \frac{3\sqrt{3}V_{peak}}{\pi}$$

If thyristors are used in place of diodes, the output voltage is reduced by a factor cos(α):

$$V_{dc} = V_{av} = \frac{3\sqrt{3}V_{peak}}{\pi}\cos\alpha$$

Or, expressed in terms of the line to line input voltage:

$$V_{dc} = V_{av} = \frac{3V_{LLpeak}}{\pi}\cos\alpha$$

Where:

V_{LLpeak}, the peak value of the line to line input voltages,

V_{peak}, the peak value of the phase (line to neutral) input voltages,

α, firing angle of the thyristor (0 if diodes are used to perform rectification)

The above equations are only valid when no current is drawn from the AC supply or in the theoretical case when the AC supply connections have no inductance. In practice, the supply inductance causes a reduction of DC output voltage with increasing load, typically in the range 10–20% at full load.

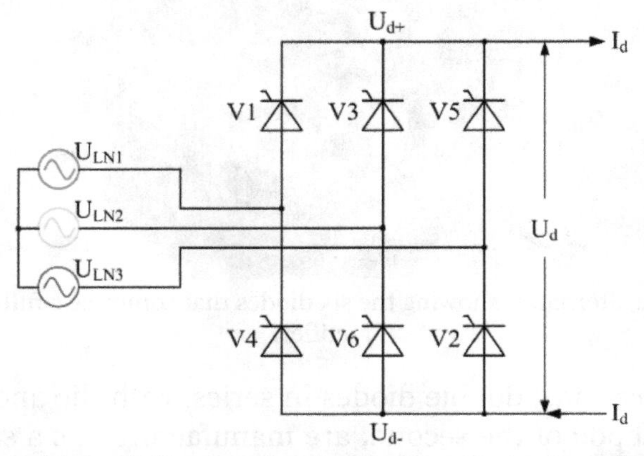

Three-phase full-wave bridge rectifier circuit using thyristors as the switching e lements, ignoring supply inductance

The effect of supply inductance is to slow down the transfer process (called commutation) from one phase to the next. As result of this is that at each transition between a pair of devices, there is a period of overlap during which three (rather than two) devices in the bridge are conducting simultaneously. The overlap angle is usually referred to by the symbol μ (or u), and may be 20 30° at full load.

With supply inductance taken into account, the output voltage of the rectifier is reduced to:

$$V_{dc} = V_{av} = \frac{3V_{LLpeak}}{\pi}\cos(\alpha + \mu)$$

The overlap angle μ is directly related to the DC current, and the above equation may be re-expressed as:

$$V_{dc} = V_{av} = \frac{3V_{LLpeak}}{\pi}\cos(\alpha) - 6fL_c I_d$$

Where:

L_c, the commutating inductance per phase

I_d, the direct current

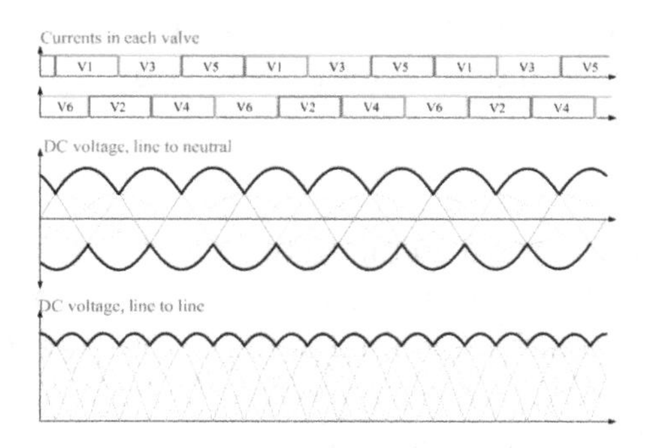

Three-phase Graetz bridge rectifier at alpha=0° without overlap

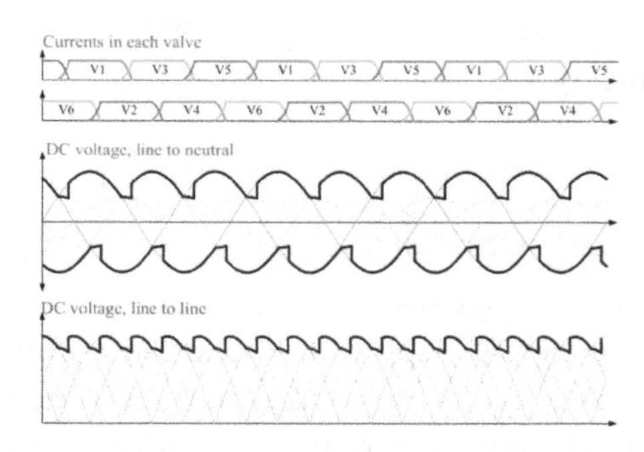

Three-phase Graetz bridge rectifier at alpha=0° with overlap angle of 20°

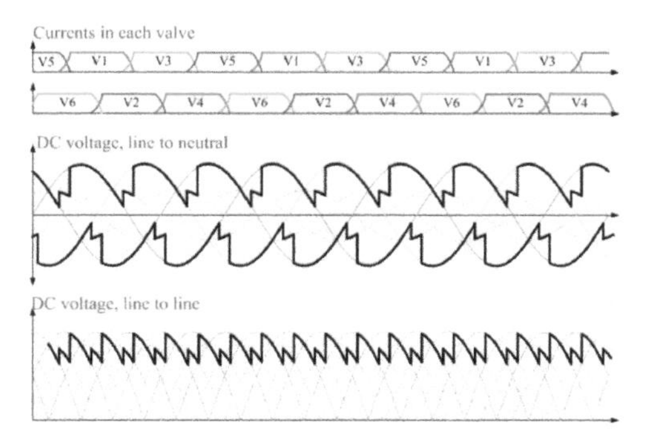

Three-phase controlled Graetz bridge rectifier at alpha=20° with overlap angle of 20°

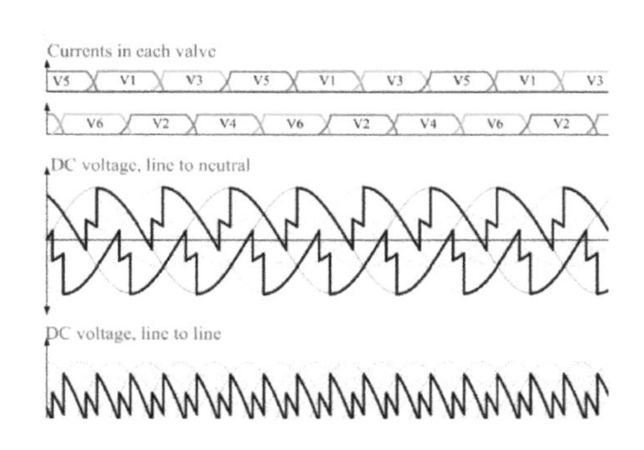

Three-phase controlled Graetz bridge rectifier at alpha=40° with overlap angle of 20°

Twelve-pulse Bridge

Although better than single-phase rectifiers or three-phase half-wave rectifiers, six-pulse rectifier circuits still produce considerable harmonic distortion on both the AC and DC connections. For very high-power rectifiers the twelve-pulse bridge connection is usually used. A twelve-pulse bridge consists of two six-pulse bridge circuits connected in series, with their AC connections fed from a supply transformer that produces a 30° phase shift between the two bridges. This cancels many of the characteristic harmonics the six-pulse bridges produce.

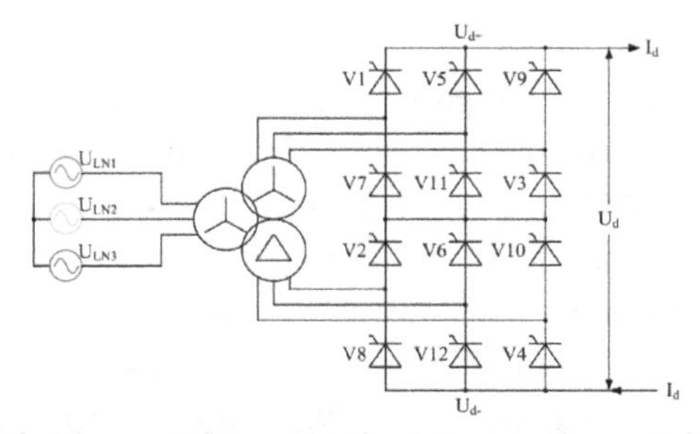

Twelve pulse bridge rectifier using thyristors as the switching elements

The 30 degree phase shift is usually achieved by using a transformer with two sets of secondary windings, one in star (wye) connection and one in delta connection.

Voltage-multiplying Rectifiers

The simple half-wave rectifier can be built in two electrical configurations with the diode pointing in opposite directions, one version connects the negative terminal of the output direct to the AC supply and the other connects the positive terminal of the output direct to the AC supply. By combining both of these with separate output smoothing it is possible to get an output voltage of nearly double the peak AC input voltage. This also provides a tap in the middle, which allows use of such a circuit as a split rail power supply.

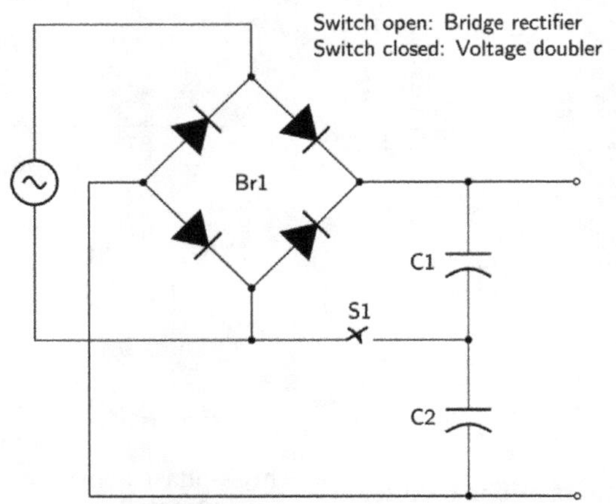

Switchable full bridge/voltage doubler.

A variant of this is to use two capacitors in series for the output smoothing on a bridge rectifier then place a switch between the midpoint of those capacitors and one of the AC input terminals. With the switch open, this circuit acts like a normal bridge rectifier. With the switch closed, it act like a voltage doubling rectifier. In other words, this makes it easy to derive a voltage of roughly 320 V (±15%, approx.) DC from any 120 V or 230 V mains supply in the world, this can then be fed into a relatively simple switched-mode power supply. However, for a given desired ripple, the value of both capacitors must be twice the value of the single one required for a normal bridge rectifier; when the switch is closed each one must filter the output of a half-wave rectifier, and when the switch is open the two capacitors are connected in series with an equivalent value of half one of them.

Cascaded diode and capacitor stages can be added to make a voltage multiplier (Cockroft-Walton circuit). These circuits are capable of producing a DC output voltage potential tens of times that of the peak AC input voltage, but are limited in current capacity and regulation. Diode voltage multipliers, frequently used as a trailing boost stage or primary high voltage (HV) source, are used in HV laser power supplies, powering devices such as cathode ray tubes (CRT) (like those used in CRT based television, radar and sonar displays), photon amplifying devices found in image intensifying and photo multiplier tubes (PMT), and magnetron based radio frequency (RF) devices used in radar transmitters and microwave ovens. Before the introduction of semiconductor electronics, transformerless powered vacuum tube receivers powered directly from AC power sometimes used voltage doublers to generate about 170 VDC from a 100–120 V power line.

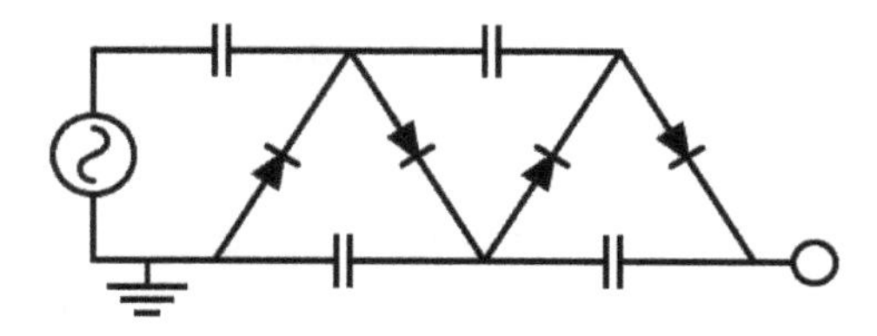

Cockcroft Walton voltage multiplier

Rectifier Efficiency

Rectifier efficiency (η) is defined as the ratio of DC output power to the input power from the AC supply. Even with ideal rectifiers with no losses, the efficiency is less than 100% because some of the output power is AC power rather than DC which manifests as ripple superimposed on the DC waveform. For a half-wave rectifier efficiency is very poor,

$$P_{in} = \frac{V_{peak}}{2} \cdot \frac{I_{peak}}{2}$$

(the divisors are 2 rather than √2 because no power is delivered on the negative half-cycle)

$$P_{out} = \frac{V_{peak}}{\pi} \cdot \frac{I_{peak}}{\pi}$$

Thus maximum efficiency for a half-wave rectifier is,

$$\eta = \frac{P_{out}}{P_{in}} = \frac{4}{\pi^2} \approx 40.5\%$$

Similarly, for a full-wave rectifier,

$$\eta = \frac{P_{out}}{P_{in}} = \frac{8}{\pi^2} \approx 81.0\%$$

Efficiency is reduced by losses in transformer windings and power dissipation in the rectifier element itself. Efficiency can be improved with the use of smoothing circuits which reduce the ripple and hence reduce the AC content of the output. Three-phase rectifiers, especially three-phase full-wave rectifiers, have much greater efficiencies because the ripple is intrinsically smaller. In some three-phase and multi-phase applications the efficiency is high enough that smoothing circuitry is unnecessary.

Rectifier Losses

A real rectifier characteristically drops part of the input voltage (a voltage drop, for silicon devices, of typically 0.7 volts plus an equivalent resistance, in general non-linear)—

and at high frequencies, distorts waveforms in other ways. Unlike an ideal rectifier, it dissipates some power.

An aspect of most rectification is a loss from the peak input voltage to the peak output voltage, caused by the built-in voltage drop across the diodes (around 0.7 V for ordinary silicon p–n junction diodes and 0.3 V for Schottky diodes). Half-wave rectification and full-wave rectification using a center-tapped secondary produces a peak voltage loss of one diode drop. Bridge rectification has a loss of two diode drops. This reduces output voltage, and limits the available output voltage if a very low alternating voltage must be rectified. As the diodes do not conduct below this voltage, the circuit only passes current through for a portion of each half-cycle, causing short segments of zero voltage (where instantaneous input voltage is below one or two diode drops) to appear between each "hump".

Peak loss is very important for low voltage rectifiers (for example, 12 V or less) but is insignificant in high-voltage applications such as HVDC.

Rectifier Output Smoothing

While half-wave and full-wave rectification can deliver unidirectional current, neither produces a constant voltage. Producing steady DC from a rectified AC supply requires a smoothing circuit or filter. In its simplest form this can be just a reservoir capacitor or smoothing capacitor, placed at the DC output of the rectifier. There is still an AC ripple voltage component at the power supply frequency for a half-wave rectifier, twice that for full-wave, where the voltage is not completely smoothed.

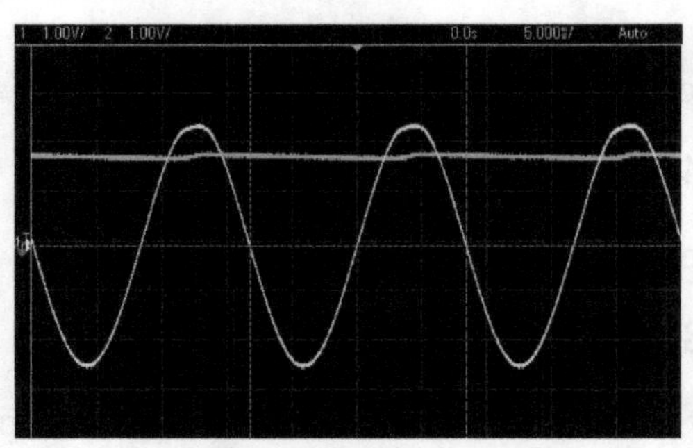

The AC input (yellow) and DC output (green) of a half-wave rectifier with a smoothing capacitor. Note the ripple in the DC signal.

Sizing of the capacitor represents a tradeoff. For a given load, a larger capacitor reduces ripple but costs more and creates higher peak currents in the transformer secondary and in the supply that feeds it. The peak current is set in principle by the rate of rise of the supply voltage on the rising edge of the incoming sine-wave, but in practice it is reduced by the resistance of the transformer windings. In extreme cases where many rectifiers are loaded onto a power distribution circuit, peak currents may cause difficulty in maintaining a correctly shaped sinusoidal voltage on the ac supply.

To limit ripple to a specified value the required capacitor size is proportional to the load current and inversely proportional to the supply frequency and the number of output peaks of the rectifier per input cycle. The load current and the supply frequency are generally outside the control of the designer of the rectifier system but the number of peaks per input cycle can be affected by the choice of rectifier design.

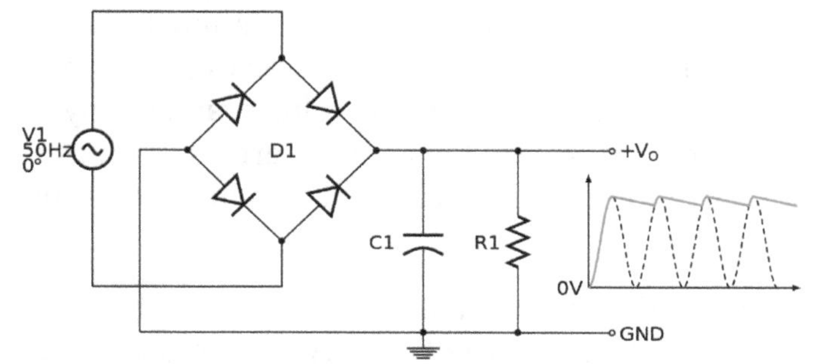

RC-Filter Rectifier: This circuit was designed and simulated using Multisim 8 software.

A half-wave rectifier only gives one peak per cycle, and for this and other reasons is only used in very small power supplies. A full wave rectifier achieves two peaks per cycle, the best possible with a single-phase input. For three-phase inputs a three-phase bridge gives six peaks per cycle. Higher numbers of peaks can be achieved by using transformer networks placed before the rectifier to convert to a higher phase order.

To further reduce ripple, a capacitor-input filter can be used. This complements the reservoir capacitor with a choke (inductor) and a second filter capacitor, so that a steadier DC output can be obtained across the terminals of the filter capacitor. The choke presents a high impedance to the ripple current. For use at power-line frequencies inductors require cores of iron or other magnetic materials, and add weight and size. Their use in power supplies for electronic equipment has therefore dwindled in favour of semiconductor circuits such as voltage regulators.

A more usual alternative to a filter, and essential if the DC load requires very low ripple voltage, is to follow the reservoir capacitor with an active voltage regulator circuit. The reservoir capacitor must be large enough to prevent the troughs of the ripple dropping below the minimum voltage required by the regulator to produce the required output voltage. The regulator serves both to significantly reduce the ripple and to deal with variations in supply and load characteristics. It would be possible to use a smaller reservoir capacitor (these can be large on high-current power supplies) and then apply some filtering as well as the regulator, but this is not a common strategy. The extreme of this approach is to dispense with the reservoir capacitor altogether and put the rectified waveform straight into a choke-input filter. The advantage of this circuit is that the current waveform is smoother and consequently the rectifier no longer has to deal with the current as a large current pulse, but instead the current delivery is spread over the entire cycle. The disadvantage, apart from extra size and weight, is that the voltage output is much lower – approximately the average of an AC half-cycle rather than the peak.

Applications

The primary application of rectifiers is to derive DC power from an AC supply (AC to DC converter). Virtually all electronic devices require DC, so rectifiers are used inside the power supplies of virtually all electronic equipment.

Converting DC power from one voltage to another is much more complicated. One method of DC-to-DC conversion first converts power to AC (using a device called an inverter), then uses a transformer to change the voltage, and finally rectifies power back to DC. A frequency of typically several tens of kilohertz is used, as this requires much smaller inductance than at lower frequencies and obviates the use of heavy, bulky, and expensive iron-cored units.

Rectifiers are also used for detection of amplitude modulated radio signals. The signal may be amplified before detection. If not, a very low voltage drop diode or a diode biased with a fixed voltage must be used. When using a rectifier for demodulation the capacitor and load resistance must be carefully matched: too low a capacitance makes the high frequency carrier pass to the output, and too high makes the capacitor just charge and stay charged.

Rectifiers supply polarised voltage for welding. In such circuits control of the output current is required; this is sometimes achieved by replacing some of the diodes in a bridge rectifier with thyristors, effectively diodes whose voltage output can be regulated by switching on and off with phase fired controllers.

Thyristors are used in various classes of railway rolling stock systems so that fine control of the traction motors can be achieved. Gate turn-off thyristors are used to produce alternating current from a DC supply, for example on the Eurostar Trains to power the three-phase traction motors.

Rectification Technologies

Electromechanical

Before about 1905 when tube type rectifiers were developed, power conversion devices were purely electro-mechanical in design. Mechanical rectification systems used some form of rotation or resonant vibration (e.g. vibrators) driven by electromagnets, which operated a switch or commutator to reverse the current.

These mechanical rectifiers were noisy and had high maintenance requirements. The moving parts had friction, which required lubrication and replacement due to wear. Opening mechanical contacts under load resulted in electrical arcs and sparks that heated and eroded the contacts. They also were not able to handle AC frequencies above several thousand cycles per second.

Synchronous Rectifier

To convert alternating into direct current in electric locomotives, a synchronous rectifier may be used. It consists of a synchronous motor driving a set of heavy-duty electrical contacts. The motor spins in time with the AC frequency and periodically reverses the connections to the load at an instant when the sinusoidal current goes through a zero-crossing. The contacts do not have to switch a large current, but they must be able to carry a large current to supply the locomotive's DC traction motors.

Vibrating Rectifier

These consisted of a resonant reed, vibrated by an alternating magnetic field created by an AC electromagnet, with contacts that reversed the direction of the current on the negative half cycles. They were used in low power devices, such as battery chargers, to rectify the low voltage produced by a step-down transformer. Another use was in battery power supplies for portable vacuum tube radios, to provide the high DC voltage for the tubes. These operated as a mechanical version of modern solid state switching inverters, with a transformer to step the battery voltage up, and a set of vibrator contacts on the transformer core, operated by its magnetic field, to repeatedly break the DC battery current to create a pulsing AC to power the transformer. Then a second set of rectifier contacts on the vibrator rectified the high AC voltage from the transformer secondary to DC.

A vibrator battery charger from 1922. It produced 6A DC at 6V to charge automobile batteries.

Motor-generator Set

A motor-generator set, or the similar rotary converter, is not strictly a rectifier as it does not actually rectify current, but rather generates DC from an AC source. In an "M-G set", the shaft of an AC motor is mechanically coupled to that of a DC generator. The DC generator produces multiphase alternating currents in its armature windings, which a commutator on the armature shaft converts into a direct current output; or a homopolar generator produces a direct current without the need for a commutator. M-G sets are useful for producing DC for railway traction motors, industrial motors and other high-current applications, and were common in many high-power D.C. uses (for example, carbon-arc lamp projectors for outdoor theaters) before high-power semiconductors became widely available.

A small motor-generator set

Electrolytic

The electrolytic rectifier was a device from the early twentieth century that is no longer used. A home-made version is illustrated in the 1913 book The Boy Mechanic but it would only be suitable for use at very low voltages because of the low breakdown voltage and the risk of electric shock. A more complex device of this kind was patented by G. W. Carpenter in 1928 (US Patent 1671970).

When two different metals are suspended in an electrolyte solution, direct current flowing one way through the solution sees less resistance than in the other direction. Electrolytic rectifiers most commonly used an aluminum anode and a lead or steel cathode, suspended in a solution of tri-ammonium ortho-phosphate.

The rectification action is due to a thin coating of aluminum hydroxide on the aluminum electrode, formed by first applying a strong current to the cell to build up the coating. The rectification process is temperature-sensitive, and for best efficiency should not operate above 86 °F (30 °C). There is also a breakdown voltage where the coating is penetrated and the cell is short-circuited. Electrochemical methods are often more fragile than mechanical methods, and can be sensitive to usage variations, which can drastically change or completely disrupt the rectification processes.

Similar electrolytic devices were used as lightning arresters around the same era by suspending many aluminium cones in a tank of tri-ammonium ortho-phosphate solution. Unlike the rectifier above, only aluminium electrodes were used, and used on A.C., there was no polarization and thus no rectifier action, but the chemistry was similar.

The modern electrolytic capacitor, an essential component of most rectifier circuit configurations was also developed from the electrolytic rectifier.

Plasma Type

Mercury-arc

A rectifier used in high-voltage direct current (HVDC) power transmission systems and industrial processing between about 1909 to 1975 is a mercury-arc rectifier or mercury-arc valve. The device is enclosed in a bulbous glass vessel or large metal tub. One electrode, the cathode, is submerged in a pool of liquid mercury at the bottom of the vessel and one or more high purity graphite electrodes, called anodes, are suspended above the pool. There may be several auxiliary electrodes to aid in starting and maintaining the arc. When an electric arc is established between the cathode pool and suspended anodes, a stream of electrons flows from the cathode to the anodes through the ionized mercury, but not the other way (in principle, this is a higher-power counterpart to flame rectification, which uses the same one-way current transmission properties of the plasma naturally present in a flame).

Early 3-phase industrial mercury vapor rectifier tube

These devices can be used at power levels of hundreds of kilowatts, and may be built to handle one to six phases of AC current. Mercury-arc rectifiers have been replaced by silicon semiconductor rectifiers and high-power thyristor circuits in the mid 1970s. The most powerful mercury-arc rectifiers ever built were installed in the Manitoba Hydro Nelson River Bipole HVDC project, with a combined rating of more than 1 GW and 450 kV.

150 kV mercury-arc valve at Manitoba Hydro power station, Radisson, Canada converted AC hydropower to DC for transmission to distant cities.

Argon Gas Electron Tube

The General Electric Tungar rectifier was an argon gas-filled electron tube device with a tungsten filament cathode and a carbon button anode. It operated similarly to the thermionic vacuum tube diode, but the gas in the tube ionized during forward conduction, giving it a much lower forward voltage drop so it could rectify lower voltages. It was used for battery chargers and similar applications from the 1920s until lower-cost metal rectifiers, and later semiconductor diodes, supplanted it. These were made up to a few hundred volts and a few amperes rating, and in some sizes strongly resembled an incandescent lamp with an additional electrode.

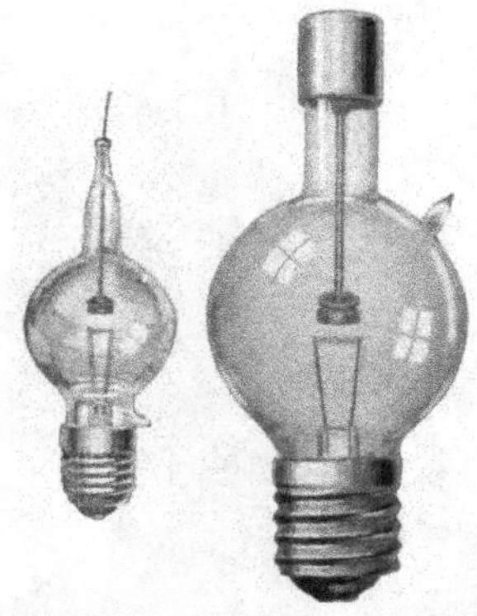

Tungar bulbs from 1917, 2 ampere (left) and 6 ampere

The 0Z4 was a gas-filled rectifier tube commonly used in vacuum tube car radios in the 1940s and 1950s. It was a conventional full-wave rectifier tube with two anodes and one cathode, but was unique in that it had no filament (thus the "0" in its type number). The electrodes were shaped such that the reverse breakdown voltage was much higher than the forward breakdown voltage. Once the breakdown voltage was exceeded, the 0Z4 switched to a low-resistance state with a forward voltage drop of about 24 V.

Diode Vacuum Tube (Valve)

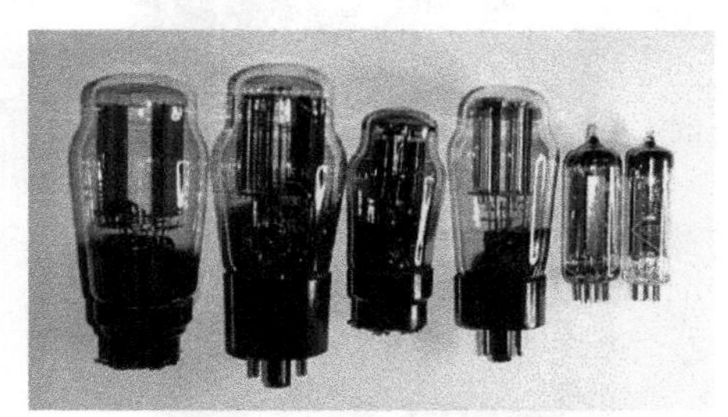

Vacuum tube diodes

The thermionic vacuum tube diode, originally called the Fleming valve, was invented by John Ambrose Fleming in 1904 as a detector for radio waves in radio receivers, and evolved into a general rectifier. It consisted of an evacuated glass bulb with a filament heated by a separate current, and a metal plate anode. The filament emitted electrons by thermionic emission (the Edison effect), discovered by Thomas Edison in 1884, and a positive voltage on the plate caused a current of electrons through the tube from filament to plate. Since only the filament produced electrons, the tube would only conduct current in one direction, allowing the tube to rectify an alternating current.

Vacuum diode rectifiers were widely used in power supplies in vacuum tube consumer electronic products, such as phonographs, radios, and televisions, for example the All American Five radio receiver, to provide the high DC plate voltage needed by other vacuum tubes. "Full-wave" versions with two separate plates were popular because they could be used with a center-tapped transformer to make a full-wave rectifier. Vacuum rectifiers were made for very high voltages, such as the high voltage power supply for the cathode ray tube of television receivers, and the kenotron used for power supply in X-ray equipment. However, compared to modern semiconductor diodes, vacuum rectifiers have high internal resistance due to space charge and therefore high voltage drops, causing high power dissipation and low efficiency. They are rarely able to handle currents exceeding 250 mA owing to the limits of plate power dissipation, and cannot be used for low voltage applications, such as battery chargers. Another limitation of the vacuum tube rectifier is that the heater power supply often requires special arrangements to insulate it from the high voltages of the rectifier circuit.

In musical instrument amplification (especially for electric guitars), the slight delay or "sag" between a signal increase (for instance, when a guitar chord is struck hard and fast) and the corresponding increase in output voltage is a notable effect of tube rectification, and results in compression. The choice between tube rectification and diode rectification is a matter of taste; some amplifiers have both and allow the player to choose.

Solid State

Crystal Detector

Galena cat's whisker detector

The cat's-whisker detector was the earliest type of semiconductor diode. It consisted of a crystal of some semiconducting mineral, usually galena (lead sulfide), with a light springy wire touching its surface. Invented by Jagadish Chandra Bose and developed by G. W. Pickard around 1906, it served as the radio wave rectifier in the first widely used radio receivers, called crystal radios. Its fragility and limited current capability made it unsuitable for power supply applications. It became obsolete around 1920, but later versions served as microwave detectors and mixers in radar receivers during World War 2.

Selenium and Copper Oxide Rectifiers

Once common until replaced by more compact and less costly silicon solid-state rectifiers in the 1970s, these units used stacks of metal plates and took advantage of the semiconductor properties of selenium or copper oxide. While selenium rectifiers were lighter in weight and used less power than comparable vacuum tube rectifiers, they had the disadvantage of finite life expectancy, increasing resistance with age, and were only suitable to use at low frequencies. Both selenium and copper oxide rectifiers have somewhat better tolerance of momentary voltage transients than silicon rectifiers.

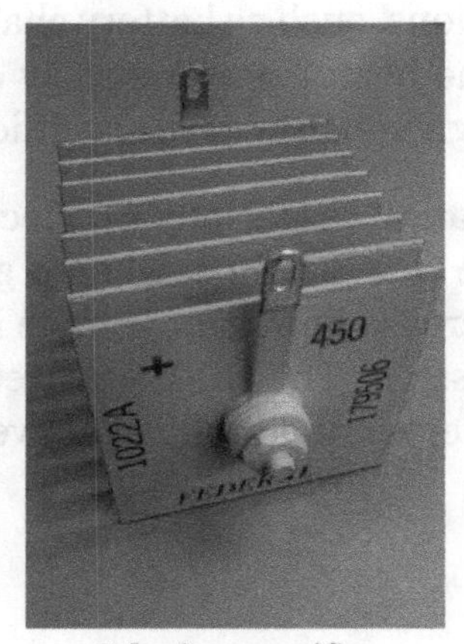

Selenium rectifier

Typically these rectifiers were made up of stacks of metal plates or washers, held together by a central bolt, with the number of stacks determined by voltage; each cell was rated for about 20 V. An automotive battery charger rectifier might have only one cell: the high-voltage power supply for a vacuum tube might have dozens of stacked plates. Current density in an air-cooled selenium stack was about 600 mA per square inch of active area (about 90 mA per square centimeter).

Silicon and Germanium Diodes

In the modern world, silicon diodes are the most widely used rectifiers for lower voltages and powers, and have largely replaced earlier germanium diodes. For very high voltages and powers, the added need for controllability has in practice led to replacing simple silicon diodes with high-power thyristors and their newer actively gate-controlled cousins.

High Power: Thyristors (SCRs) and Newer Silicon-based Voltage Sourced Converters

In high-power applications, from 1975 to 2000, most mercury valve arc-rectifiers were replaced by stacks of very high power thyristors, silicon devices with two extra layers of semiconductor, in comparison to a simple diode.

Two of three high-power thyristor valve stacks used for long distance transmission of power from Manitoba Hydro dams. Compare with mercury-arc system from the same dam-site, above.

In medium-power transmission applications, even more complex and sophisticated voltage sourced converter (VSC) silicon semiconductor rectifier systems, such as insulated gate bipolar transistors (IGBT) and gate turn-off thyristors (GTO), have made smaller high voltage DC power transmission systems economical. All of these devices function as rectifiers.

As of 2009 it was expected that these high-power silicon "self-commutating switches", in particular IGBTs and a variant thyristor (related to the GTO) called the integrated gate-commutated thyristor (IGCT), would be scaled-up in power rating to the point that they would eventually replace simple thyristor-based AC rectification systems for the highest power-transmission DC applications.

Current Research

A major area of research is to develop higher frequency rectifiers, that can rectify into terahertz and light frequencies. These devices are used in optical heterodyne detection, which has myriad applications in optical fiber communication and atomic clocks. Another prospective application for such devices is to directly rectify light waves picked up by tiny antenna, called nantennas, to produce DC electric power. It is thought that arrays of nantennas could be a more efficient means of producing solar power than solar cells.

A related area of research is to develop smaller rectifiers, because a smaller device has a higher cutoff frequency. Research projects are attempting to develop a unimolecular rectifier, a single organic molecule that would function as a rectifier.

Power Inverter

An inverter on a free-standing solar plant

A power inverter, or inverter, is an electronic device or circuitry that changes direct current (DC) to alternating current (AC).

Overview of solar-plant inverters

The input voltage, output voltage and frequency, and overall power handling depend on the design of the specific device or circuitry. The inverter does not produce any power; the power is provided by the DC source.

A power inverter can be entirely electronic or may be a combination of mechanical effects (such as a rotary apparatus) and electronic circuitry. Static inverters do not use moving parts in the conversion process.

Input and output

Input voltage

A typical power inverter device or circuit requires a relatively stable DC power source capable of supplying enough current for the intended power demands of the system. The input voltage depends on the design and purpose of the inverter. Examples include:

- 12 VDC, for smaller consumer and commercial inverters that typically run from a rechargeable 12 V lead acid battery or automotive electrical outlet.
- 24, 36 and 48 VDC, which are common standards for home energy systems.
- 200 to 400 VDC, when power is from photovoltaic solar panels.
- 300 to 450 VDC, when power is from electric vehicle battery packs in vehicle-to-grid systems.
- Hundreds of thousands of volts, where the inverter is part of a high voltage direct current power transmission system.

Output Waveform

An inverter can produce a square wave, modified sine wave, pulsed sine wave, pulse width modulated wave (PWM) or sine wave depending on circuit design. The two dominant commercialized waveform types of inverters as of 2007 are modified sine wave and sine wave.

There are two basic designs for producing household plug-in voltage from a lower-voltage DC source, the first of which uses a switching boost converter to produce a higher-voltage DC and then converts to AC. The second method converts DC to AC at battery level and uses a line-frequency transformer to create the output voltage.

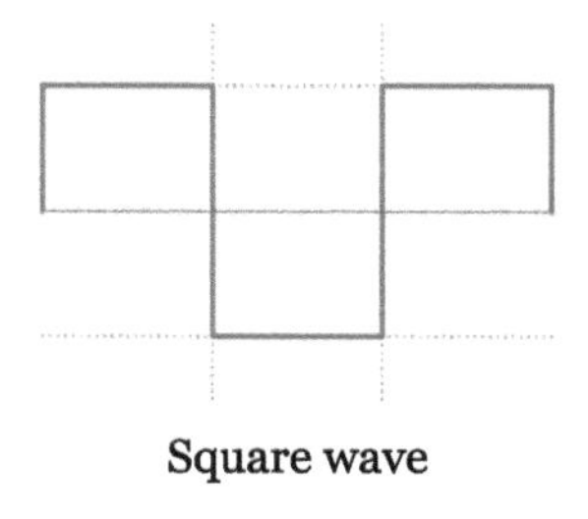

Square wave

Square wave

This is one of the simplest waveforms an inverter design can produce and is best suited to low-sensitivity applications such as lighting and heating. Square wave output can produce "humming" when connected to audio equipment and is generally unsuitable for sensitive electronics.

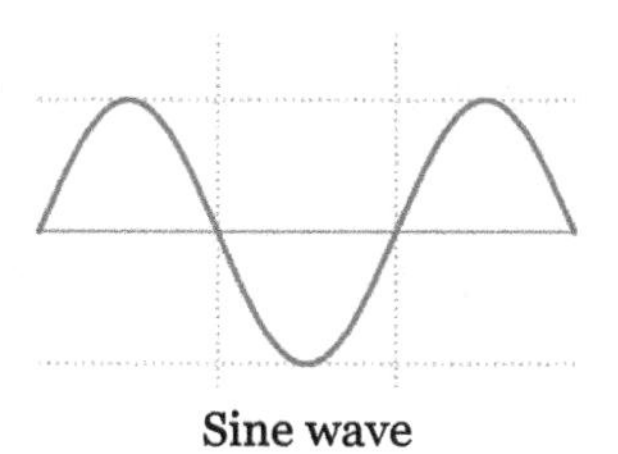

Sine wave

Sine Wave

A power inverter device which produces a multiple step sinusoidal AC waveform is referred to as a sine wave inverter. To more clearly distinguish the inverters with outputs of much less distortion than the "modified sine wave" (three step) inverter designs, the manufacturers often use the phrase pure sine wave inverter. Almost all consumer grade inverters that are sold as a "pure sine wave inverter" do not produce a smooth sine wave output at all, just a less choppy output than the square wave (two step) and modified sine wave (three step) inverters. In this sense, the phrases "Pure sine wave" or "sine wave inverter" are misleading to the consumer. However, this is not critical for most electronics as they deal with the output quite well.

Where power inverter devices substitute for standard line power, a sine wave output is desirable because many electrical products are engineered to work best with a sine wave AC power source. The standard electric utility power is a sine wave.

Sine wave inverters with more than three steps in the wave output are more complex and have significantly higher cost than a modified sine wave, with only three steps, or square wave (one step) types of the same power handling. Switch-mode power supply (SMPS) devices, such as personal computers or DVD players, function on quality modified sine wave power. AC motors directly operated on non-sinusoidal power may produce extra heat, may have different speed-torque characteristics, or may produce more audible noise than when running on sinusoidal power.

Modified Sine Wave

A modified sine wave inverter has a non-square waveform that is a useful approximation of a sine wave for power translation purposes.

Most inexpensive consumer power inverters produce a modified sine wave rather than a pure sine wave.

The waveform in commercially available modified-sine-wave inverters is a square wave with a pause before the polarity reversal, which only needs to cycle back and forth through a three-position switch that outputs forward, off, and reverse output at the pre-determined frequency. Switching states are developed for positive, negative and zero voltages as per the patterns given in the switching Table 2. The peak voltage to RMS voltage ratio does not maintain the same relationship as for a sine wave. The DC bus voltage may be actively regulated, or the "on" and "off" times can be modified to maintain the same RMS value output up to the DC bus voltage to compensate for DC bus voltage variations.

The ratio of on to off time can be adjusted to vary the RMS voltage while maintaining a constant frequency with a technique called pulse width modulation (PWM). The generated gate pulses are given to each switch in accordance with the developed pattern to obtain the desired output. Harmonic spectrum in the output depends on the width of the pulses and the modulation frequency. When operating induction motors, voltage harmonics are usually not of concern; however, harmonic distortion in the current waveform introduces additional heating and can produce pulsating torques.

Numerous items of electric equipment will operate quite well on modified sine wave power inverter devices, especially loads that are resistive in nature such as traditional incandescent light bulbs.

However, the load may operate less efficiently owing to the harmonics associated with a modified sine wave and produce a humming noise during operation. This also affects the efficiency of the system as a whole, since the manufacturer's nominal conversion efficiency does not account for harmonics. Therefore, pure sine wave inverters may provide significantly higher efficiency than modified sine wave inverters.

Most AC motors will run on MSW inverters with an efficiency reduction of about 20% owing to the harmonic content. However, they may be quite noisy. A series LC filter tuned to the fundamental frequency may help.

A common modified sine wave inverter topology found in consumer power inverters is as follows:

An onboard microcontroller rapidly switches on and off power MOSFETs at high frequency like ~50 kHz. The MOSFETs directly pull from a low voltage DC source (such as a battery). This signal then goes through step-up transformers (generally many smaller transformers are placed in parallel to reduce the overall size of the inverter) to produce a higher voltage signal. The output of the step-up transformers then gets filtered by capacitors to produce a high voltage DC supply. Finally, this DC supply is pulsed with additional power MOSFETs by the microcontroller to produce the final modified sine wave signal.

Other Waveforms

By definition there is no restriction on the type of AC waveform an inverter might produce that would find use in a specific or special application.

Output Frequency

The AC output frequency of a power inverter device is usually the same as standard power line frequency, 50 or 60 hertz

If the output of the device or circuit is to be further conditioned (for example stepped up) then the frequency may be much higher for good transformer efficiency.

Output Voltage

The AC output voltage of a power inverter is often regulated to be the same as the grid line voltage, typically 120 or 240 VAC, even when there are changes in the load that the inverter is driving. This allows the inverter to power numerous devices designed for standard line power.

Some inverters also allow selectable or continuously variable output voltages.

Output Power

A power inverter will often have an overall power rating expressed in watts or kilowatts. This describes the power that will be available to the device the inverter is driving and, indirectly, the power that will be needed from the DC source. Smaller popular consumer and commercial devices designed to mimic line power typically range from 150 to 3000 watts.

Not all inverter applications are solely or primarily concerned with power delivery; in some cases the frequency and or waveform properties are used by the follow-on circuit or device.

Batteriest

The runtime of an inverter is dependent on the battery power and the amount of power being drawn from the inverter at a given time. As the amount of equipment using the inverter increases, the runtime will decrease. In order to prolong the runtime of an inverter, additional batteries can be added to the inverter.

When attempting to add more batteries to an inverter, there are two basic options for installation:

Series configuration

> If the goal is to increase the overall voltage of the inverter, one can daisy chain batteries in a series configuration. In a series configuration, if a single battery dies, the other batteries will not be able to power the load.

Parallel configuration

> If the goal is to increase capacity and prolong the runtime of the inverter, batteries can be connected in parallel. This increases the overall ampere-hour (Ah) rating of the battery set.
>
> If a single battery is discharged though, the other batteries will then discharge through it. This can lead to rapid discharge of the entire pack, or even an over-current and possible fire. To avoid this, large paralleled batteries may be

connected via diodes or intelligent monitoring with automatic switching to isolate an under-voltage battery from the others.

Applications

DC Power Source Usage

An inverter converts the DC electricity from sources such as batteries or fuel cells to AC electricity. The electricity can be at any required voltage; in particular it can operate AC equipment designed for mains operation, or rectified to produce DC at any desired voltage.

Inverter designed to provide 115 VAC from the 12 VDC source provided in an automobile. The unit shown provides up to 1.2 amperes of alternating current, or enough to power two sixty watt light bulbs.

Uninterruptible Power Supplies

An uninterruptible power supply (UPS) uses batteries and an inverter to supply AC power when mains power is not available. When mains power is restored, a rectifier supplies DC power to recharge the batteries.

Electric Motor Speed Control

Inverter circuits designed to produce a variable output voltage range are often used within motor speed controllers. The DC power for the inverter section can be derived from a normal AC wall outlet or some other source. Control and feedback circuitry is used to adjust the final output of the inverter section which will ultimately determine the speed of the motor operating under its mechanical load. Motor speed control needs are numerous and include things like: industrial motor driven equipment, electric vehicles, rail transport systems, and power tools. Switching states are developed for positive, negative and zero voltages as per the patterns given in the switching.The generated gate pulses are given to each switch in accordance with the developed pattern and thus the output is obtained.

Power Grid

Grid-tied inverters are designed to feed into the electric power distribution system. They transfer synchronously with the line and have as little harmonic content as possible. They also need a means of detecting the presence of utility power for safety reasons, so as not to continue to dangerously feed power to the grid during a power outage.

Solar

A solar inverter is a balance of system (BOS) component of a photovoltaic system and can be used for both, grid-connected and off-grid systems. Solar inverters have special functions adapted for use with photovoltaic arrays, including maximum power point tracking and anti-islanding protection. Solar micro-inverters differ from conventional converters, as an individual micro-converter is attached to each solar panel. This can improve the overall efficiency of the system. The output from several microinverters is then combined and often fed to the electrical grid.

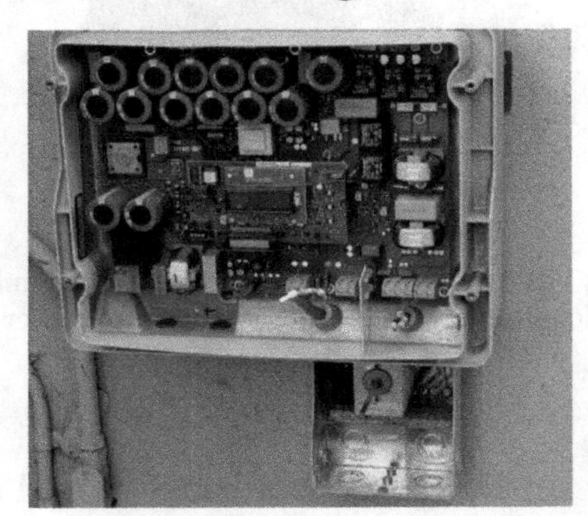

Internal view of a solar inverter. Note the many large capacitors (blue cylinders), used to store energy briefly and improve the output waveform.

Induction Heating

Inverters convert low frequency main AC power to higher frequency for use in induction heating. To do this, AC power is first rectified to provide DC power. The inverter then changes the DC power to high frequency AC power. Due to the reduction in the number of DC sources employed, the structure becomes more reliable and the output voltage has higher resolution due to an increase in the number of steps so that the reference sinusoidal voltage can be better achieved. This configuration has recently become very popular in AC power supply and adjustable speed drive applications. This new inverter can avoid extra clamping diodes or voltage balancing capacitors.

There are three kinds of level shifted modulation techniques, namely:

- Phase Opposition Disposition (POD)
- Alternative Phase Opposition Disposition (APOD)
- Phase Disposition (PD)

HVDC Power Transmission

With HVDC power transmission, AC power is rectified and high voltage DC power is transmitted to another location. At the receiving location, an inverter in a static inverter plant converts the power back to AC. The inverter must be synchronized with grid frequency and phase and minimize harmonic generation.

The High voltage DC transmission method can be useful for things like Solar power since solar power is natively DC as it is.

Electroshock Weapons

Electroshock weapons and tasers have a DC/AC inverter to generate several tens of thousands of V AC out of a small 9 V DC battery. First the 9 V DC is converted to 400–2000 V AC with a compact high frequency transformer, which is then rectified and temporarily stored in a high voltage capacitor until a pre-set threshold voltage is reached. When the threshold (set by way of an airgap or TRIAC) is reached, the capacitor dumps its entire load into a pulse transformer which then steps it up to its final output voltage of 20–60 kV. A variant of the principle is also used in electronic flash and bug zappers, though they rely on a capacitor-based voltage multiplier to achieve their high voltage.

Miscellaneous

Typical applications for power inverters include:

- Portable consumer devices that allow the user to connect a battery, or set of batteries, to the device to produce AC power to run various electrical items such as lights, televisions, kitchen appliances, and power tools.
- Use in power generation systems such as electric utility companies or solar generating systems to convert DC power to AC power.
- Use within any larger electronic system where an engineering need exists for deriving an AC source from a DC source.

Circuit Description

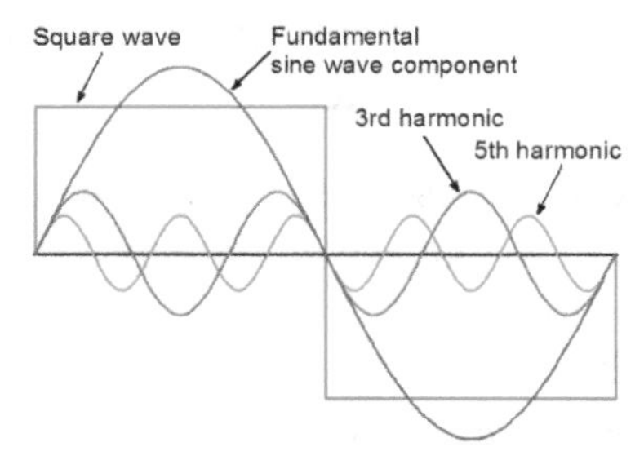

Square waveform with fundamental sine wave component, 3rd harmonic and 5th harmonic

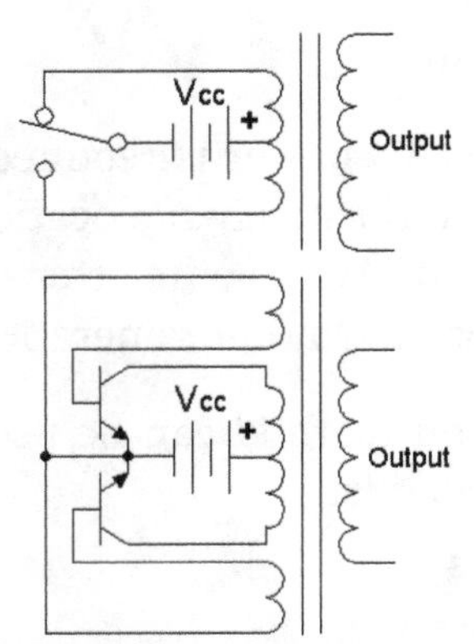

Top: Simple inverter circuit shown with an electromechanical switch and automatic equivalent auto-switching device implemented with two transistors and split winding auto-transformer in place of the mechanical switch.

Basic Design

In one simple inverter circuit, DC power is connected to a transformer through the center tap of the primary winding. A switch is rapidly switched back and forth to allow current to flow back to the DC source following two alternate paths through one end of the primary winding and then the other. The alternation of the direction of current in the primary winding of the transformer produces alternating current (AC) in the secondary circuit.

The electromechanical version of the switching device includes two stationary contacts and a spring supported moving contact. The spring holds the movable contact against one of the stationary contacts and an electromagnet pulls the movable contact to the opposite stationary contact. The current in the electromagnet is interrupted by the action of the switch so that the switch continually switches rapidly back and forth. This type of electromechanical inverter switch, called a vibrator or buzzer, was once used in vacuum tube automobile radios. A similar mechanism has been used in door bells, buzzers and tattoo machines.

As they became available with adequate power ratings, transistors and various other types of semiconductor switches have been incorporated into inverter circuit designs. Certain ratings, especially for large systems (many kilowatts) use thyristors (SCR). SCRs provide large power handling capability in a semiconductor device, and can readily be controlled over a variable firing range.

The switch in the simple inverter described above, when not coupled to an output transformer, produces a square voltage waveform due to its simple off and on nature as opposed to the sinusoidal waveform that is the usual waveform of an AC power supply. Using Fourier analysis, periodic waveforms are represented as the sum of an infinite series of sine waves. The sine wave that has the same frequency as the original waveform is called the fundamental component. The other sine waves, called harmonics, that are included in the series have frequencies that are integral multiples of the fundamental frequency.

Fourier analysis can be used to calculate the total harmonic distortion (THD). The total harmonic distortion (THD) is the square root of the sum of the squares of the harmonic voltages divided by the fundamental voltage:

$$\mathrm{THD} = \frac{\sqrt{V_2^2 + V_3^2 + V_4^2 + \cdots + V_n^2}}{V_1}$$

Advanced Designs

There are many different power circuit topologies and control strategies used in inverter designs. Different design approaches address various issues that may be more or less important depending on the way that the inverter is intended to be used.

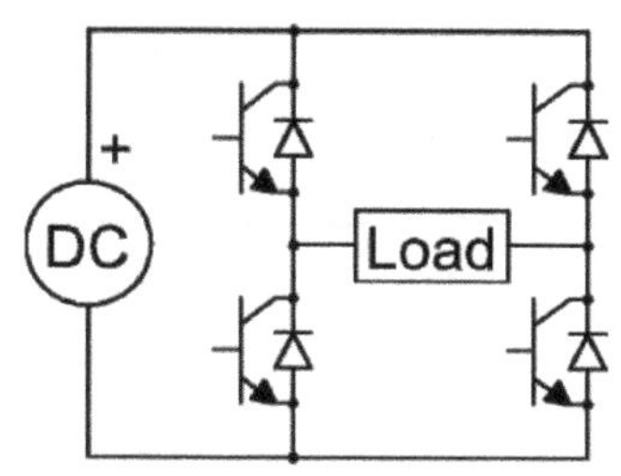

H bridge inverter circuit with transistor switches and antiparallel diodes

The issue of waveform quality can be addressed in many ways. Capacitors and inductors can be used to filter the waveform. If the design includes a transformer, filtering can be applied to the primary or the secondary side of the transformer or to both sides. Low-pass filters are applied to allow the fundamental component of the waveform to pass to the output while limiting the passage of the harmonic components. If the inverter is designed to provide power at a fixed frequency, a resonant filter can be used. For an adjustable frequency inverter, the filter must be tuned to a frequency that is above the maximum fundamental frequency.

Since most loads contain inductance, feedback rectifiers or antiparallel diodes are often connected across each semiconductor switch to provide a path for the peak inductive load current when the switch is turned off. The antiparallel diodes are somewhat similar to the freewheeling diodes used in AC/DC converter circuits.

Waveform	Signal transitions per period	Harmonics eliminated	Harmonics amplified	System description	THD
	2			2-level square wave	~45%

	4	3, 9, 27, ...		3-level "modified square wave"	>23.8%
	8			5-level "modified square wave"	>6.5%
	10	3, 5, 9, 27	7, 11, ...	2-level very slow PWM	
	12	3, 5, 9, 27	7, 11, ...	3-level very slow PWM	

Fourier analysis reveals that a waveform, like a square wave, that is anti-symmetrical about the 180 degree point contains only odd harmonics, the 3rd, 5th, 7th, etc. Waveforms that have steps of certain widths and heights can attenuate certain lower harmonics at the expense of amplifying higher harmonics. For example, by inserting a zero-voltage step between the positive and negative sections of the square-wave, all of the harmonics that are divisible by three (3rd and 9th, etc.) can be eliminated. That leaves only the 5th, 7th, 11th, 13th etc. The required width of the steps is one third of the period for each of the positive and negative steps and one sixth of the period for each of the zero-voltage steps.

Changing the square wave as described above is an example of pulse-width modulation (PWM). Modulating, or regulating the width of a square-wave pulse is often used as a method of regulating or adjusting an inverter's output voltage. When voltage control is not required, a fixed pulse width can be selected to reduce or eliminate selected harmonics. Harmonic elimination techniques are generally applied to the lowest harmonics because filtering is much more practical at high frequencies, where the filter components can be much smaller and less expensive. Multiple pulse-width or carrier based PWM control schemes produce waveforms that are composed of many narrow pulses. The frequency represented by the number of narrow pulses per second is called the switching frequency or carrier frequency. These control schemes are often used in variable-frequency motor control inverters because they allow a wide range of output voltage and frequency adjustment while also improving the quality of the waveform.

Multilevel inverters provide another approach to harmonic cancellation. Multilevel inverters provide an output waveform that exhibits multiple steps at several voltage levels. For example, it is possible to produce a more sinusoidal wave by having split-rail direct current inputs at two voltages, or positive and negative inputs with a central ground. By connecting the inverter output terminals in sequence between the positive rail and ground, the positive rail and the negative rail, the ground rail and the negative rail, then both to the ground rail, a stepped waveform is generated at the inverter output. This is an example of a three level inverter: the two voltages and ground.

More on Achieving A Sine Wave

Resonant inverters produce sine waves with LC circuits to remove the harmonics from a simple square wave. Typically there are several series- and parallel-resonant LC circuits, each tuned to a different harmonic of the power line frequency. This simplifies the electronics, but the inductors and capacitors tend to be large and heavy. Its high efficiency makes this approach popular in large uninterruptible power supplies in data centers that run the inverter continuously in an "online" mode to avoid any switchover transient when power is lost.

A closely related approach uses a ferroresonant transformer, also known as a constant voltage transformer, to remove harmonics and to store enough energy to sustain the load for a few AC cycles. This property makes them useful in standby power supplies to eliminate the switchover transient that otherwise occurs during a power failure while the normally idle inverter starts and the mechanical relays are switching to its output.

Enhanced Quantization

A proposal suggested in Power Electronics magazine utilizes two voltages as an improvement over the common commercialized technology which can only apply DC bus voltage in either directions or turn it off. The proposal adds an additional voltage to this design. Each cycle consists of sequence as: v1, v2, v1, 0, −v1, −v2, −v1.

Three-phase Inverters

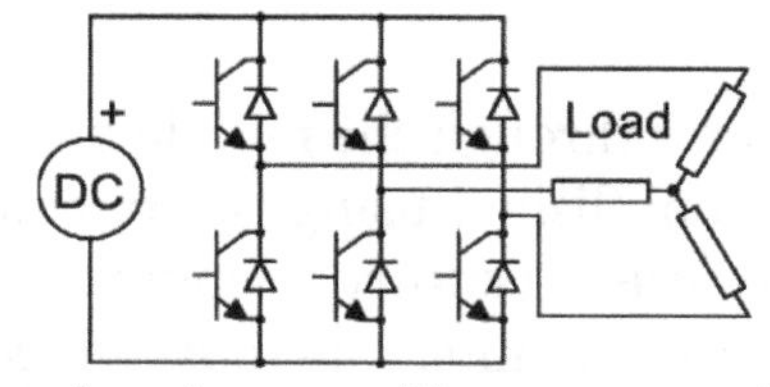

Three-phase inverter with wye connected load

Three-phase inverters are used for variable-frequency drive applications and for high power applications such as HVDC power transmission. A basic three-phase inverter consists of three single-phase inverter switches each connected to one of the three load terminals. For the most basic control scheme, the operation of the three switches is coordinated so that one switch operates at each 60 degree point of the fundamental output waveform. This creates a line-to-line output waveform that has six steps. The six-step waveform has a zero-voltage step between the positive and negative sections of the square-wave such that the harmonics that are multiples of three are eliminated as described above. When carrier-based PWM techniques are applied to six-step waveforms, the basic overall shape, or envelope, of the waveform is retained so that the 3rd harmonic and its multiples are cancelled.

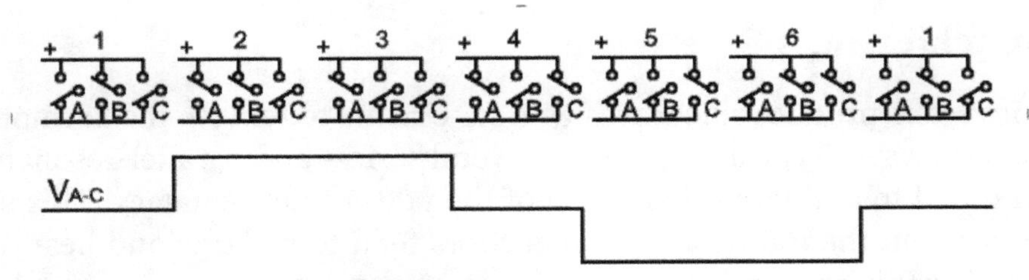

3-phase inverter switching circuit showing 6-step switching sequence and waveform of voltage between terminals A and C ($2^3 - 2$ states)

To construct inverters with higher power ratings, two six-step three-phase inverters can be connected in parallel for a higher current rating or in series for a higher voltage rating. In either case, the output waveforms are phase shifted to obtain a 12-step waveform. If additional inverters are combined, an 18-step inverter is obtained with three inverters etc. Although inverters are usually combined for the purpose of achieving increased voltage or current ratings, the quality of the waveform is improved as well.

Size

Compared to other household electric devices, inverters are large in size and volume. In 2014 Google together with IEEE started an open competition to build a (much) smaller power inverter, with a $1,000,000 prize.

History

Early Inverters

From the late nineteenth century through the middle of the twentieth century, DC-to-AC power conversion was accomplished using rotary converters or motor-generator sets (M-G sets). In the early twentieth century, vacuum tubes and gas filled tubes began to be used as switches in inverter circuits. The most widely used type of tube was the thyratron.

The origins of electromechanical inverters explain the source of the term inverter. Early AC-to-DC converters used an induction or synchronous AC motor direct-connected to a generator (dynamo) so that the generator's commutator reversed its connections at exactly the right moments to produce DC. A later development is the synchronous converter, in which the motor and generator windings are combined into one armature, with slip rings at one end and a commutator at the other and only one field frame. The result with either is AC-in, DC-out. With an M-G set, the DC can be considered to be separately generated from the AC; with a synchronous converter, in a certain sense it can be considered to be "mechanically rectified AC". Given the right auxiliary and control equipment, an M-G set or rotary converter can be "run backwards", converting DC to AC. Hence an inverter is an inverted converter.

Controlled Rectifier Inverters

Since early transistors were not available with sufficient voltage and current ratings for most inverter applications, it was the 1957 introduction of the thyristor or silicon-controlled rectifier (SCR) that initiated the transition to solid state inverter circuits.

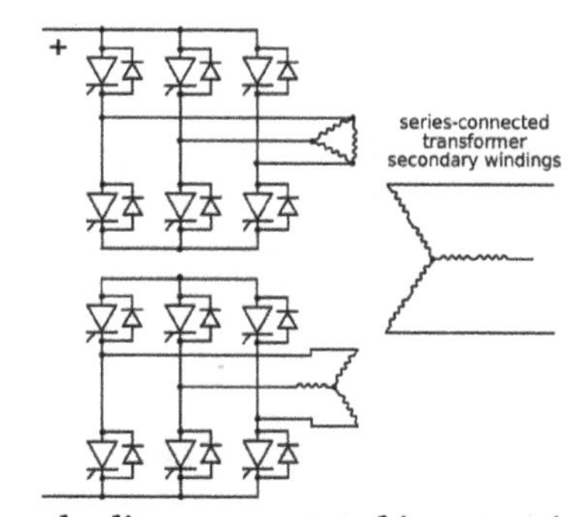

12-pulse line-commutated inverter circuit

The commutation requirements of SCRs are a key consideration in SCR circuit designs. SCRs do not turn off or commutate automatically when the gate control signal is shut off. They only turn off when the forward current is reduced to below the minimum holding current, which varies with each kind of SCR, through some external process. For SCRs connected to an AC power source, commutation occurs naturally every time the polarity of the source voltage reverses. SCRs connected to a DC power source usually require a means of forced commutation that forces the current to zero when commutation is required. The least complicated SCR circuits employ natural commutation rather than forced commutation. With the addition of forced commutation circuits, SCRs have been used in the types of inverter circuits described above.

In applications where inverters transfer power from a DC power source to an AC power source, it is possible to use AC-to-DC controlled rectifier circuits operating in the inversion mode. In the inversion mode, a controlled rectifier circuit operates as a line commutated inverter. This type of operation can be used in HVDC power transmission systems and in regenerative braking operation of motor control systems.

Another type of SCR inverter circuit is the current source input (CSI) inverter. A CSI inverter is the dual of a six-step voltage source inverter. With a current source inverter, the DC power supply is configured as a current source rather than a voltage source. The inverter SCRs are switched in a six-step sequence to direct the current to a three-phase AC load as a stepped current waveform. CSI inverter commutation methods include load commutation and parallel capacitor commutation. With both methods, the input current regulation assists the commutation. With load commutation, the load is a synchronous motor operated at a leading power factor.

As they have become available in higher voltage and current ratings, semiconductors such as transistors or IGBTs that can be turned off by means of control signals have become the preferred switching components for use in inverter circuits.

Rectifier and Inverter Pulse Numbers

Rectifier circuits are often classified by the number of current pulses that flow to the DC side of the rectifier per cycle of AC input voltage. A single-phase half-wave rectifier is a one-pulse circuit and a single-phase full-wave rectifier is a two-pulse circuit. A three-phase half-wave rectifier is a three-pulse circuit and a three-phase full-wave rectifier is a six-pulse circuit.

With three-phase rectifiers, two or more rectifiers are sometimes connected in series or parallel to obtain higher voltage or current ratings. The rectifier inputs are supplied from special transformers that provide phase shifted outputs. This has the effect of phase multiplication. Six phases are obtained from two transformers, twelve phases from three transformers and so on. The associated rectifier circuits are 12-pulse rectifi ers, 18-pulse rectifiers and so on...

When controlled rectifier circuits are operated in the inversion mode, they would be classified by pulse number also. Rectifier circuits that have a higher pulse number have reduced harmonic content in the AC input current and reduced ripple in the DC output voltage. In the inversion mode, circuits that have a higher pulse number have lower harmonic content in the AC output voltage waveform.

Research

Using 3-D printing and novel semiconductors, researchers at the Department of Energy's Oak Ridge National Laboratory have created a power inverter that could make electric vehicles lighter, more powerful and more efficient.

References

- Stephen Sangwine (2 March 2007). Electronic Components and Technology, Third Edition. CRC Press. p. 73. ISBN 978-1-4200-0768-8.
- Lander, Cyril W. (1993). "2. Rectifying Circuits". Power electronics (3rd ed.). London: McGraw-Hill. ISBN 978-0-07-707714-3.
- Williams, B. W. (1992). "Chapter 11". Power electronics : devices, drivers and applications (2nd ed.). Basingstoke: Macmillan. ISBN 978-0-333-57351-8.
- Wendy Middleton, Mac E. Van Valkenburg (eds), Reference Data for Engineers: Radio, Electronics, Computer, and Communications, p. 14. 13, Newnes, 2002 ISBN 0-7506-7291-9.
- Arrillaga, Jos; Liu, Yonghe H; Watson, Neville R; Murray, Nicholas J. Self-Commutating Converters for High Power Applications. John Wiley & Sons. ISBN 978-0-470-68212-8.
- Kimbark, Edward Wilson (1971). Direct current transmission. (4. printing. ed.). New York: Wiley-Interscience. p. 508. ISBN 978-0-471-47580-4.
- Jeff Barrow of Integrated Device Technology, Inc. (21 November 2011). "Understand and reduce DC/DC switching-converter ground noise". Eetimes.com. Retrieved 18 January 2016.
- Hendrik Rissik (1941). Mercury-arc current convertors [sic] : an introduction to the theory and practice of vapour-arc discharge devices and to the study of rectification phenomena. Sir I. Pitman & sons, ltd. Retrieved 8 January 2013.
- Hawkins, Nehemiah (1914). "54. Rectifiers". Hawkins Electrical Guide: Principles of electricity, magnetism, induction, experiments, dynamo. New York: T. Audel. Retrieved 8 January 2013.
- American Technical Society (1920). Cyclopedia of applied electricity. 2. American technical society. p. 487. Retrieved 8 January 2013.

3

Methods of Generating Electricity

The generation of electricity from sources like wind using wind turbines, sunlight using photovoltaic devices and magnets using electromagnetic induction has been explored in this content. The principle and methodology of these methods of generating electricity have been discussed in-depth. There is a section devoted to static electricity, its causes and applications as well.

Static Electricity

Static electricity is an imbalance of electric charges within or on the surface of a material. The charge remains until it is able to move away by means of an electric current or electrical discharge. Static electricity is named in contrast with current electricity, which flows through wires or other conductors and transmits energy.

Contact with the slide has left this child's hair positively charged so that the individual hairs repel one another. The hair can also be attracted to the negatively charged slide surface.

A static electric charge can be created whenever two surfaces contact and separate, and at least one of the surfaces has a high resistance to electric current (and is therefore an electrical insulator). The effects of static electricity are familiar to most people because people can feel, hear, and even see the spark as the excess charge is neutralized when brought close to a large electrical conductor (for example, a path to ground), or a region with an excess charge of the opposite polarity (positive or negative). The familiar phenomenon of a static shock–more specifically, an electrostatic discharge–is caused by the neutralization of charge.

Causes of Static Electricity

Materials are made of atoms that are normally electrically neutral because they contain equal numbers of positive charges (protons in their nuclei) and negative charges (electrons in "shells" surrounding the nucleus). The phenomenon of static electricity requires a separation of positive and negative charges. When two materials are in contact, electrons may move from one material to the other, which leaves an excess of positive charge on one material, and an equal negative charge on the other. When the materials are separated they retain this charge imbalance.

Contact-induced Charge Separation

Electrons can be exchanged between materials on contact; materials with weakly bound electrons tend to lose them while materials with sparsely filled outer shells tend to gain them. This is known as the triboelectric effect and results in one material becoming positively charged and the other negatively charged. The polarity and strength of the charge on a material once they are separated depends on their relative positions in the triboelectric series. The triboelectric effect is the main cause of static electricity as observed in everyday life, and in common high-school science demonstrations involving rubbing different materials together (e.g., fur against an acrylic rod). Contact-induced charge separation causes your hair to stand up and causes "static cling" (for example, a balloon rubbed against the hair becomes negatively charged; when near a wall, the charged balloon is attracted to positively charged particles in the wall, and can "cling" to it, appearing to be suspended against gravity).

Pressure-induced Charge Separation

Applied mechanical stress generates a separation of charge in certain types of crystals and ceramics molecules.

Heat-induced Charge Separation

Heating generates a separation of charge in the atoms or molecules of certain materials. All pyroelectric materials are also piezoelectric. The atomic or molecular properties of heat and pressure response are closely related.

Charge-induced Charge Separation

A charged object brought close to an electrically neutral object causes a separation of charge within the neutral object. Charges of the same polarity are repelled and charges of the opposite polarity are attracted. As the force due to the interaction of electric charges falls off rapidly with increasing distance, the effect of the closer (opposite polarity) charges is greater and the two objects feel a force of attraction. The effect is most pronounced when the neutral object is an electrical conductor as the charges are more free to move around. Careful grounding of part of an object with a charge-induced charge separation can permanently add or remove electrons, leaving the object with a global, permanent charge. This process is integral to the workings of the Van de Graaff generator, a device commonly used to demonstrate the effects of static electricity.

Removal and Prevention of Static Electricity

Removing or preventing a buildup of static charge can be as simple as opening a window or using a humidifier to increase the moisture content of the air, making the atmosphere more conductive. Air ionizers can perform the same task.

Items that are particularly sensitive to static discharge may be treated with the application of an antistatic agent, which adds a conducting surface layer that ensures any excess charge is evenly distributed. Fabric softeners and dryer sheets used in washing machines and clothes dryers are an example of an antistatic agent used to prevent and remove static cling.

Many semiconductor devices used in electronics are particularly sensitive to static discharge. Conductive antistatic bags are commonly used to protect such components. People who work on circuits that contain these devices often ground themselves with a conductive antistatic strap.

In the industrial settings such as paint or flour plants as well as in hospitals, antistatic safety boots are sometimes used to prevent a buildup of static charge due to contact with the floor. These shoes have soles with good conductivity. Anti-static shoes should not be confused with insulating shoes, which provide exactly the opposite benefit – some protection against serious electric shocks from the mains voltage.

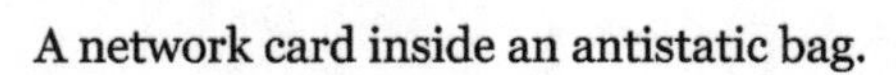

A network card inside an antistatic bag.

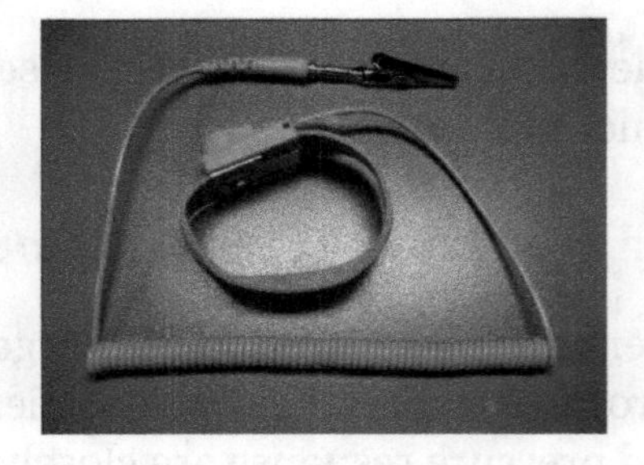

An antistatic wrist strap with crocodile clip.

Static Discharge

The spark associated with static electricity is caused by electrostatic discharge, or simply static discharge, as excess charge is neutralized by a flow of charges from or to the surroundings.

The feeling of an electric shock is caused by the stimulation of nerves as the neutralizing current flows through the human body. The energy stored as static electricity on an object varies depending on the size of the object and its capacitance, the voltage to which it is charged, and the dielectric constant of the surrounding medium. For modelling the effect of static discharge on sensitive electronic devices, a human being is represented as a capacitor of 100 picofarads, charged to a voltage of 4000 to 35000 volts. When touching an object this energy is discharged in less than a microsecond. While the total energy is small, on the order of millijoules, it can still damage sensitive electronic devices. Larger objects will store more energy, which may be directly hazardous to human contact or which may give a spark that can ignite flammable gas or dust.

Lightning

Lightning is a dramatic natural example of static discharge. While the details are unclear and remain a subject of debate, the initial charge separation is thought to be associated with contact between ice particles within storm clouds. In general, significant charge accumulations can only persist in regions of low electrical conductivity (very few charges free to move in the surroundings), hence the flow of neutralizing charges often results from neutral atoms and molecules in the air being torn apart to form separate positive and negative charges, which travel in opposite directions as an electric current, neutralizing the original accumulation of charge. The static charge in air typically breaks down in this way at around 10,000 volts per centimeter (10 kV/cm) depending on humidity. The discharge superheats the surrounding air causing the bright flash, and produces a shock wave causing the clicking sound. The lightning bolt is simply a scaled-up version of the sparks seen in more domestic occurrences of static discharge. The flash occurs because the air in the discharge channel is heated to such a high temperature that it emits light by incandescence. The clap of thunder is the result of the shock wave created as the superheated air expands explosively.

Natural static discharge

Electronic Components

Many semiconductor devices used in electronics are very sensitive to the presence of static electricity and can be damaged by a static discharge. The use of an antistatic strap is mandatory for researchers manipulating nanodevices. Further precautions can be taken by taking off shoes with thick rubber soles and permanently staying with a metallic ground.

Static Build-up in Flowing Flammable and Ignitable Materials

Discharge of static electricity can create severe hazards in those industries dealing with flammable substances, where a small electrical spark might ignite explosive mixtures.

The flowing movement of finely powdered substances or low conductivity fluids in pipes or through mechanical agitation can build up static electricity. The flow of granules of material like sand down a plastic chute can transfer charge, which can be easily measured using a multimeter connected to metal foil lining the chute at intervals, and can be roughly proportional to particulate flow. Dust clouds of finely powdered substances can become combustible or explosive. When there is a static discharge in a dust or vapor cloud, explosions have occurred. Among the major industrial incidents that have occurred are: a grain silo in southwest France, a paint plant in Thailand, a factory making fiberglass moldings in Canada, a storage tank explosion in Glenpool, Oklahoma in 2003, and a portable tank filling operation and a tank farm in Des Moines, Iowa and Valley Center, Kansas in 2007.

The ability of a fluid to retain an electrostatic charge depends on its electrical conductivity. When low conductivity fluids flow through pipelines or are mechanically agitated, contact-induced charge separation called flow electrification occurs. Fluids that have low electrical conductivity (below 50 picosiemens per meter), are called accumulators. Fluids having conductivities above 50 pS/m are called non-accumulators. In non-accumulators, charges recombine as fast as they are separated and hence electrostatic charge accumulation is not significant. In the petrochemical industry, 50 pS/m is the recommended minimum value of electrical conductivity for adequate removal of charge from a fluid.

Kerosines may have conductivity ranging from less than 1 picosiemens per meter to 20 pS/m. For comparison, deionized water has a conductivity of about 10,000,000 pS/m or 10 μS/m.

Transformer oil is part of the electrical insulation system of large power transformers and other electrical apparatus. Re-filling of large apparatus requires precautions against electrostatic charging of the fluid, which may damage sensitive transformer insulation.

An important concept for insulating fluids is the static relaxation time. This is similar to the time constant τ (tau) within an RC circuit. For insulating materials, it is the ratio of the static dielectric constant divided by the electrical conductivity of the material. For hydrocarbon fluids, this is sometimes approximated by dividing the number 18 by the electrical conductivity of the fluid. Thus a fluid that has an electrical conductivity of 1 pS/m has an estimated relaxation time of about 18 seconds. The excess charge in a fluid dissipates almost completely after four to five times the relaxation time, or 90 seconds for the fluid in the above example.

Charge generation increases at higher fluid velocities and larger pipe diameters, becoming quite significant in pipes 8 inches (200 mm) or larger. Static charge generation in these systems is best controlled by limiting fluid velocity. The British standard BS PD CLC/TR 50404:2003 (formerly BS-5958-Part 2) Code of Practice for Control of Undesirable Static Electricity prescribes pipe flow velocity limits. Because water content has a large impact on the fluids dielectric constant, the recommended velocity for hydrocarbon fluids containing water should be limited to 1 meter per second.

Bonding and earthing are the usual ways charge buildup can be prevented. For fluids with electrical conductivity below 10 pS/m, bonding and earthing are not adequate for charge dissipation, and anti-static additives may be required.

Fueling Operations

The flowing movement of flammable liquids like gasoline inside a pipe can build up static electricity. Non-polar liquids such as gasoline, toluene, xylene, diesel, kerosene and light crude oils exhibit significant ability for charge accumulation and charge retention during high velocity flow. Electrostatic discharges can ignite the fuel vapor. When the electrostatic discharge energy is high enough, it can ignite a fuel vapor and air mixture. Different fuels have different flammable limits and require different levels of electrostatic discharge energy to ignite.

Electrostatic discharge while fueling with gasoline is a present danger at gas stations. Fires have also been started at airports while refueling aircraft with kerosene. New grounding technologies, the use of conducting materials, and the addition of anti-static additives help to prevent or safely dissipate the buildup of static electricity.

The flowing movement of gases in pipes alone creates little, if any, static electricity. It is envisaged that a charge generation mechanism only occurs when solid particles or liquid droplets are carried in the gas stream.

Static Discharge in Space Exploration

Due to the extremely low humidity in extraterrestrial environments, very large static charges can accumulate, causing a major hazard for the complex electronics used in space exploration vehicles. Static electricity is thought to be a particular hazard for astronauts on planned missions to the Moon and Mars. Walking over the extremely dry terrain could cause them to accumulate a significant amount of charge; reaching out to open the airlock on their return could cause a large static discharge, potentially damaging sensitive electronics.

Ozone Cracking

A static discharge in the presence of air or oxygen can create ozone. Ozone can degrade rubber parts. Many elastomers are sensitive to ozone cracking. Exposure to ozone creates deep penetrative cracks in critical components like gaskets and O-rings. Fuel lines are also susceptible to the problem unless preventive action is taken. Preventive measures include adding anti-ozonants to the rubber mix, or using an ozone-resistant elastomer. Fires from cracked fuel lines have been a problem on vehicles, especially in the engine compartments where ozone can be produced by electrical equipment.

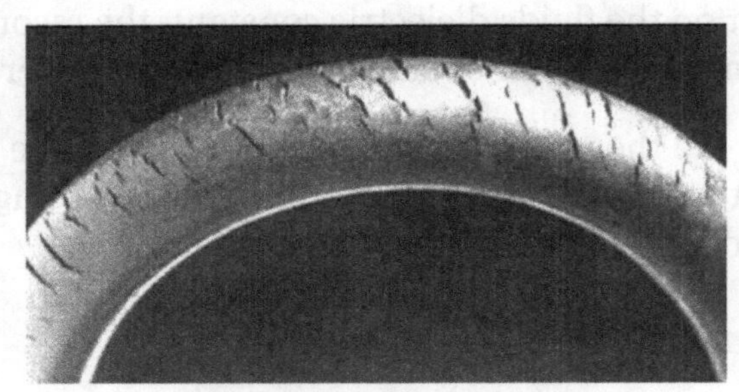

Ozone cracking in natural rubber tubing

Energies Involved

The energy released in a static electricity discharge may vary over a wide range. The energy in joules can be calculated from the capacitance (C) of the object and the static potential V in volts (V) by the formula $E = \frac{1}{2}CV^2$. One experimenter estimates the capacitance of the human body as high as 400 picofarads, and a charge of 50,000 volts, discharged e.g. during touching a charged car, creating a spark with energy of 500 millijoules. Another estimate is 100–300 pF and 20,000 volts, producing a maximum energy of 60 mJ. IEC 479-2:1987 states that a discharge with energy greater than 5000 mJ is a direct serious risk to human health. IEC 60065 states that consumer products cannot discharge more than 350 mJ into a person.

The maximum potential is limited to about 35–40 kV, due to corona discharge dissipating the charge at higher potentials. Potentials below 3000 volts are not typically detectable by humans. Maximum potential commonly achieved on human body range between 1 and 10 kV, though in optimal conditions as high as 20–25 kV can be reached. Low relative humidity increases the charge buildup; walking 20 feet (6.1 m) on vinyl floor at 15% relative humidity causes buildup of voltage up to 12 kilovolts, while at 80% humidity the voltage is only 1.5 kV.

As little as 0.2 millijoules may present an ignition hazard; such low spark energy is often below the threshold of human visual and auditory perception.

Typical ignition energies are:

- 0.017 mJ for hydrogen
- 0.2-2 mJ for hydrocarbon vapors
- 1–50 mJ for fine flammable dust

- 40–1000 mJ for coarse flammable dust.

The energy needed to damage most electronic devices is between 2 and 1000 nanojoules.

A relatively small energy, often as little as 0.2–2 millijoules, is needed to ignite a flammable mixture of a fuel and air. For the common industrial hydrocarbon gases and solvents, the minimum ignition energy required for ignition of vapor-air mixture is lowest for the vapor concentration roughly in the middle between the lower explosive limit and the upper explosive limit, and rapidly increases as the concentration deviates from this optimum to either side. Aerosols of flammable liquids may be ignited well below their flash point. Generally, liquid aerosols with particle sizes below 10 micrometers behave like vapors, particle sizes above 40 micrometers behave more like flammable dusts. Typical minimum flammable concentrations of aerosols lay between 15 and 50 g/m^3. Similarly, presence of foam on the surface of a flammable liquid significantly increases ignitability. Aerosol of flammable dust can be ignited as well, resulting in a dust explosion; the lower explosive limit usually lies between 50 and 1000 g/m^3; finer dusts tend to be more explosive and requiring less spark energy to set off. Simultaneous presence of flammable vapors and flammable dust can significantly decrease the ignition energy; a mere 1 vol.% of propane in air can reduce the required ignition energy of dust by 100 times. Higher than normal oxygen content in atmosphere also significantly lowers the ignition energy.

There are five types of electrical discharges:

- Spark, responsible for the majority of industrial fires and explosions where static electricity is involved. Sparks occur between objects at different electric potentials. Good grounding of all parts of the equipment and precautions against charge buildups on equipment and personnel are used as prevention measures.
- Brush discharge occurs from a nonconductive charged surface or highly charged nonconductive liquids. The energy is limited to roughly 4 millijoules. To be hazardous, the voltage involved must be above about 20 kilovolts, the surface polarity must be negative, a flammable atmosphere must be present at the point of discharge, and the discharge energy must be sufficient for ignition. Further, because surfaces have a maximum charge density, an area of at least 100 cm^2 has to be involved. This is not considered to be a hazard for dust clouds.
- Propagating brush discharge is high in energy and dangerous. Occurs when an insulating surface of up to 8 mm thick (e.g. a teflon or glass lining of a grounded metal pipe or a reactor) is subjected to a large charge buildup between the opposite surfaces, acting as a large-area capacitor.
- Cone discharge, also called bulking brush discharge, occurs over surfaces of charged powders with resistivity above 10^{10} ohms, or also deep through the pow-

der mass. Cone discharges aren't usually observed in dust volumes below 1 m^3. The energy involved depends on the grain size of the powder and the charge magnitude, and can reach up to 20 mJ. Larger dust volumes produce higher energies.

- Corona discharge, considered non-hazardous.

Applications of Static Electricity

Static electricity is commonly used in xerography, air filters (particularly electrostatic precipitators), automotive paints, photocopiers, paint sprayers, theaters, flooring in operating theaters, powder testing, printers, static bonding and aircraft refueling.

Wind Turbine

Offshore wind farm, using 5 MW turbines REpower 5M in the North Sea off the coast of Belgium.

A wind turbine is a device that converts the wind's kinetic energy into electrical power.

Wind turbines are manufactured in a wide range of vertical and horizontal axis types. The smallest turbines are used for applications such as battery charging for auxiliary power for boats or caravans or to power traffic warning signs. Slightly larger turbines can be used for making contributions to a domestic power supply while selling unused power back to the utility supplier via the electrical grid. Arrays of large turbines, known as wind farms, are becoming an increasingly important source of intermittent renewable energy and are used by many countries as part of a strategy to reduce their reliance on fossil fuels.

History

Windmills were used in Persia (present-day Iran) about 500-900 A.D. The windwheel of Hero of Alexandria marks one of the first known instances of wind powering a machine in history. However, the first known practical windmills were built in Sistan, an Eastern province of Iran, from the 7th century. These "Panemone" were vertical axle windmills, which had long vertical drive shafts with rectangular blades. Made of six to twelve sails covered in reed matting or cloth material, these windmills were used to grind grain or draw up water, and were used in the gristmilling and sugarcane industries.

James Blyth's electricity-generating wind turbine, photographed in 1891

Windmills first appeared in Europe during the Middle Ages. The first historical records of their use in England date to the 11th or 12th centuries and there are reports of German crusaders taking their windmill-making skills to Syria around 1190. By the 14th century, Dutch windmills were in use to drain areas of the Rhine delta. Advanced wind mills were described by Croatian inventor Fausto Veranzio. In his book Machinae Novae (1595) he described vertical axis wind turbines with curved or V-shaped blades.

The first electricity-generating wind turbine was a battery charging machine installed in July 1887 by Scottish academic James Blyth to light his holiday home in Marykirk, Scotland. Some months later American inventor Charles F. Brush was able to build the first automatically operated wind turbine after consulting local University professors and colleagues Jacob S. Gibbs and Brinsley Coleberd and successfully getting the blueprints peer-reviewed for electricity production in Cleveland, Ohio. Although Blyth's turbine was considered uneconomical in the United Kingdom electricity generation by wind turbines was more cost effective in countries with widely scattered populations.

In Denmark by 1900, there were about 2500 windmills for mechanical loads such as pumps and mills, producing an estimated combined peak power of about 30 MW. The largest machines were on 24-meter (79 ft) towers with four-bladed 23-meter (75 ft) diameter rotors. By 1908 there were 72 wind-driven electric generators operating in the United States from 5 kW to 25 kW. Around the time of World War I, American windmill makers were producing 100,000 farm windmills each year, mostly for water-pumping.

By the 1930s, wind generators for electricity were common on farms, mostly in the United States where distribution systems had not yet been installed. In this period, high-tensile steel was cheap, and the generators were placed atop prefabricated open steel lattice towers.

The first automatically operated wind turbine, built in Cleveland in 1887 by Charles F. Brush. It was 60 feet (18 m) tall, weighed 4 tons (3.6 metric tonnes) and powered a 12 kW generator.

A forerunner of modern horizontal-axis wind generators was in service at Yalta, USSR in 1931. This was a 100 kW generator on a 30-meter (98 ft) tower, connected to the local 6.3 kV distribution system. It was reported to have an annual capacity factor of 32 percent, not much different from current wind machines.

In the autumn of 1941, the first megawatt-class wind turbine was synchronized to a utility grid in Vermont. The Smith-Putnam wind turbine only ran for 1,100 hours before suffering a critical failure. The unit was not repaired, because of shortage of materials during the war.

The first utility grid-connected wind turbine to operate in the UK was built by John Brown & Company in 1951 in the Orkney Islands.

Despite these diverse developments, developments in fossil fuel systems almost entirely eliminated any wind turbine systems larger than supermicro size. In the early 1970s, however, anti-nuclear protests in Denmark spurred artisan mechanics to develop microturbines of 22 kW. Organizing owners into associations and co-operatives lead to the lobbying of the government and utilities and provided incentives for larger turbines throughout the 1980s and later. Local activists in Germany, nascent turbine manufacturers in Spain, and large investors in the United States in the early 1990s then lobbied for policies that stimulated the industry in those countries. Later companies formed in India and China. As of 2012, Danish company Vestas is the world's biggest wind-turbine manufacturer.

Resources

A quantitative measure of the wind energy available at any location is called the Wind Power Density (WPD). It is a calculation of the mean annual power available per square meter of swept area of a turbine, and is tabulated for different heights above ground. Calculation of wind power density includes the effect of wind velocity and air density. Color-coded maps are prepared for a particular area described, for example, as "Mean Annual Power Density at 50 Metres". In the United States, the results of the above calculation are included in an index developed by the National Renewable Energy Laboratory and referred to as "NREL CLASS". The larger the WPD, the higher it is rated by class. Classes range from Class 1 (200 watts per square meter or less at 50 m altitude) to Class 7 (800 to 2000 watts per square m). Commercial wind farms generally are sited in Class 3 or higher areas, although isolated points in an otherwise Class 1 area may be practical to exploit.

Nordex N117/2400 in Germany, a modern low-wind turbine.

Wind turbines are classified by the wind speed they are designed for, from class I to class IV, with A or B referring to the turbulence.

Class	Avg Wind Speed (m/s)	Turbulence
IA	10	18%
IB	10	16%
IIA	8.5	18%
IIB	8.5	16%
IIIA	7.5	18%
IIIB	7.5	16%
IVA	6	18%
IVB	6	16%

Efficiency

Not all the energy of blowing wind can be used, but some small wind turbines are designed to work at low wind speeds.

Conservation of mass requires that the amount of air entering and exiting a turbine must be equal. Accordingly, Betz's law gives the maximal achievable extraction of wind power by a wind turbine as 16/27 (59.3%) of the total kinetic energy of the air flowing through the turbine.

The maximum theoretical power output of a wind machine is thus 0.59 times the kinetic energy of the air passing through the effective disk area of the machine. If the effective area of the disk is A, and the wind velocity v, the maximum theoretical power output P is:

$$P = 0.59\frac{1}{2}\rho v^3 A$$

where ρ is air density

As wind is free (no fuel cost), wind-to-rotor efficiency (including rotor blade friction and drag) is one of many aspects impacting the final price of wind power. Further inefficiencies, such as gearbox losses, generator and converter losses, reduce the power delivered by a wind turbine. To protect components from undue wear, extracted power is held constant above the rated operating speed as theoretical power increases at the cube of wind speed, further reducing theoretical efficiency. In 2001, commercial utility-connected turbines deliver 75% to 80% of the Betz limit of power extractable from the wind, at rated operating speed.

Efficiency can decrease slightly over time due to wear. Analysis of 3128 wind turbines older than 10 years in Denmark showed that half of the turbines had no decrease, while the other half saw a production decrease of 1.2% per year. Vertical turbine designs have much lower efficiency than standard horizontal designs.

Types

Wind turbines can rotate about either a horizontal or a vertical axis, the former being both older and more common. They can also include blades (transparent or not) or be bladeless. Vertical designs produce less power and are less common.

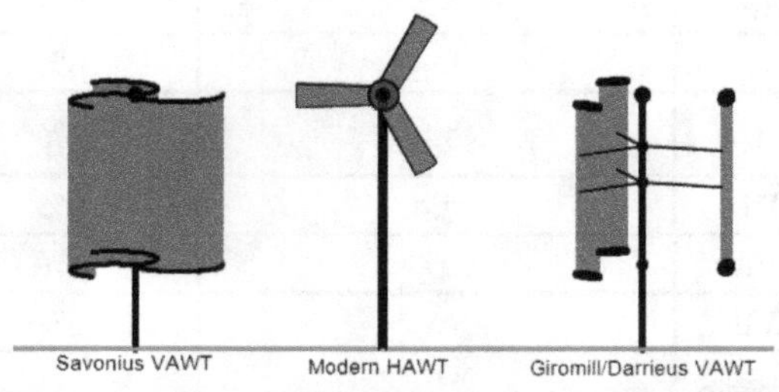

The three primary types: VAWT Savonius, HAWT towered; VAWT Darrieus as they appear in operation

Horizontal Axis

Horizontal-axis wind turbines (HAWT) have the main rotor shaft and electrical generator at the top of a tower, and must be pointed into the wind. Small turbines are pointed by a simple wind vane, while large turbines generally use a wind sensor coupled with a servo motor. Most have a gearbox, which turns the slow rotation of the blades into a quicker rotation that is more suitable to drive an electrical generator.

Components of a horizontal axis wind turbine (gearbox, rotor shaft and brake assembly) being lifted into position

Since a tower produces turbulence behind it, the turbine is usually positioned upwind of its supporting tower. Turbine blades are made stiff to prevent the blades from being pushed into the tower by high winds. Additionally, the blades are placed a considerable distance in front of the tower and are sometimes tilted forward into the wind a small amount.

A turbine blade convoy passing through Edenfield, UK

Downwind machines have been built, despite the problem of turbulence (mast wake), because they don't need an additional mechanism for keeping them in line with the wind, and because in high winds the blades can be allowed to bend which reduces their swept area and thus their wind resistance. Since cyclical (that is repetitive) turbulence may lead to fatigue failures, most HAWTs are of upwind design.

Turbines used in wind farms for commercial production of electric power are usually three-bladed and pointed into the wind by computer-controlled motors. These have high tip speeds of over 320 km/h (200 mph), high efficiency, and low torque ripple, which contribute to good reliability. The blades are usually colored white for daytime visibility by aircraft and range in length from 20 to 40 meters (66 to 131 ft) or more. The tubular steel towers range from 60 to 90 meters (200 to 300 ft) tall.

The blades rotate at 10 to 22 revolutions per minute. At 22 rotations per minute the tip speed exceeds 90 meters per second (300 ft/s). A gear box is commonly used for stepping up the speed of the generator, although designs may also use direct drive of an annular generator. Some models operate at constant speed, but more energy can be collected by variable-speed turbines which use a solid-state power converter to interface to the transmission system. All turbines are equipped with protective features to avoid damage at high wind speeds, by feathering the blades into the wind which ceases their rotation, supplemented by brakes.

Year by year the size and height of turbines increase. Offshore wind turbines are built up to 8MW today and have a blade length up to 80m. Onshore wind turbines are installed in low wind speed areas and getting higher and higher towers. Usual towers of multi megawatt turbines have a height of 70 m to 120 m and in extremes up to 160 m, with blade tip speeds reaching 80 m/s to 90 m/s. Higher tip speeds means more noise and blade erosion.

Vertical Axis Design

Vertical-axis wind turbines (or VAWTs) have the main rotor shaft arranged vertically. One advantage of this arrangement is that the turbine does not need to be pointed into the wind to be effective, which is an advantage on a site where the wind direction is highly variable. It is also an advantage when the turbine is integrated into a building because it is inherently less steerable. Also, the generator and gearbox can be placed near the ground, using a direct drive from the rotor assembly to the ground-based gearbox, improving accessibility for maintenance. However, these designs produce much less energy averaged over time, which is a major drawback.

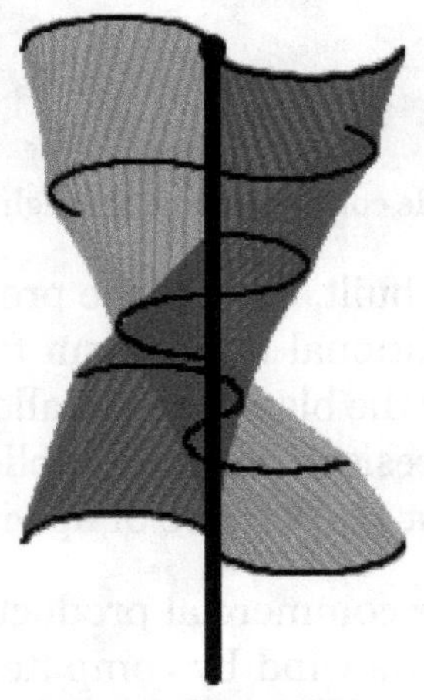

A vertical axis Twisted Savonius type turbine.

The key disadvantages include the relatively low rotational speed with the consequential higher torque and hence higher cost of the drive train, the inherently lower power coefficient, the 360-degree rotation of the aerofoil within the wind flow during each cycle and hence the highly dynamic loading on the blade, the pulsating torque generated by some rotor designs on the drive train, and the difficulty of modelling the wind flow accurately and hence the challenges of analysing and designing the rotor prior to fabricating a prototype.

When a turbine is mounted on a rooftop the building generally redirects wind over the roof and this can double the wind speed at the turbine. If the height of a rooftop mounted turbine tower is approximately 50% of the building height it is near the optimum for maximum wind energy and minimum wind turbulence. Wind speeds within the built environment are generally much lower than at exposed rural sites, noise may be a concern and an existing structure may not adequately resist the additional stress.

Offshore Horizontal Axis Wind Turbines (HAWTs) at Scroby Sands Wind Farm, UK

Subtypes of the vertical axis design include:

Onshore Horizontal Axis Wind Turbines in Zhangjiakou, China

Darrieus wind turbine

> "Eggbeater" turbines, or Darrieus turbines, were named after the French inventor, Georges Darrieus. They have good efficiency, but produce large torque ripple and cyclical stress on the tower, which contributes to poor reliability. They also generally require some external power source, or an additional Savonius rotor to start turning, because the starting torque is very low. The torque ripple is reduced by using three or more blades which results in greater solidity of the rotor. Solidity is measured by blade area divided by the rotor area. Newer Darrieus type turbines are not held up by guy-wires but have an external superstructure connected to the top bearing.

Giromill

A subtype of Darrieus turbine with straight, as opposed to curved, blades. The cycloturbine variety has variable pitch to reduce the torque pulsation and is self-starting. The advantages of variable pitch are: high starting torque; a wide, relatively flat torque curve; a higher coefficient of performance; more efficient operation in turbulent winds; and a lower blade speed ratio which lowers blade bending stresses. Straight, V, or curved blades may be used.

Savonius Wind Turbine

These are drag-type devices with two (or more) scoops that are used in anemometers, Flettner vents (commonly seen on bus and van roofs), and in some high-reliability low-efficiency power turbines. They are always self-starting if there are at least three scoops.

Twisted Savonius

Twisted Savonius is a modified savonius, with long helical scoops to provide smooth torque. This is often used as a rooftop windturbine and has even been adapted for ships.

Another type of vertical axis is the Parallel turbine, which is similar to the crossflow fan or centrifugal fan. It uses the ground effect. Vertical axis turbines of this type have been tried for many years: a unit producing 10 kW was built by Israeli wind pioneer Bruce Brill in the 1980s.

Vortexis

The most recent advancement in Vertical Axis Wind Turbines has been the Vortexis VAWT, utilizing a pre-swirled augmented vertical axis wind turbine (PA-VAWT) designed for the purpose of developing a high efficiency VAWT concept that keeps the advantages of VAWT's compact size, lack of bias as to incoming wind direction, easy deployment and low radar cross section for use in mobile applications for the military, referred to in Special Operations as "Black Swan."

Design and Construction

Wind turbines are designed to exploit the wind energy that exists at a location. Aerodynamic modeling is used to determine the optimum tower height, control systems, number of blades and blade shape.

Wind turbines convert wind energy to electricity for distribution. Conventional horizontal axis turbines can be divided into three components:

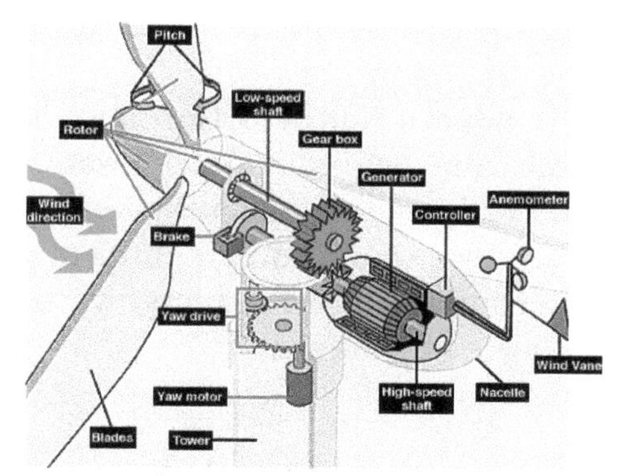

Components of a horizontal-axis wind turbine

Inside view of a wind turbine tower, showing the tendon cables.

- The rotor component, which is approximately 20% of the wind turbine cost, includes the blades for converting wind energy to low speed rotational energy.
- The generator component, which is approximately 34% of the wind turbine cost, includes the electrical generator, the control electronics, and most likely a gearbox (e.g. planetary gearbox), adjustable-speed drive or continuously variable transmission component for converting the low speed incoming rotation to high speed rotation suitable for generating electricity.
- The structural support component, which is approximately 15% of the wind turbine cost, includes the tower and rotor yaw mechanism.

A 1.5 MW wind turbine of a type frequently seen in the United States has a tower 80 meters (260 ft) high. The rotor assembly (blades and hub) weighs 22,000 kilograms (48,000 lb). The nacelle, which contains the generator component, weighs 52,000 kilograms (115,000 lb). The concrete base for the tower is constructed using 26,000 kilograms (58,000 lb) of reinforcing steel and contains 190 cubic meters (250 cu yd) of concrete. The base is 15 meters (50 ft) in diameter and 2.4 meters (8 ft) thick near the center.

Among all renewable energy systems wind turbines have the highest effective intensity of power-harvesting surface because turbine blades not only harvest wind power, but also concentrate it.

Unconventional Designs

An E-66 wind turbine in the Windpark Holtriem, Germany, has an observation deck for visitors. Another turbine of the same type with an observation deck is located in Swaffham, England. Airborne wind turbine designs have been proposed and developed for many years but have yet to produce significant amounts of energy. In principle, wind turbines may also be used in conjunction with a large vertical solar updraft tower to extract the energy due to air heated by the sun.

Counter rotating wind turbine (dual rotor)

A ram air turbine (RAT) is a special kind of small turbine that is fitted to some aircraft. When deployed, the RAT is spun by the airstream going past the aircraft and can provide power for the most essential systems if there is a loss of all on-board electrical power, as in the case of the "Gimli Glider".

The corkscrew shaped wind turbine at Progressive Field in Cleveland, Ohio

Wind turbines which utilise the Magnus effect have been developed.

The two-bladed turbine SCD 6MW offshore turbine designed by aerodyn Energiesysteme, built by MingYang Wind Power has a helideck for helicopters on top of its nacelle. The prototype was erected in 2014 in Rudong China.

Turbine Monitoring and Diagnostics

Due to data transmission problems, structural health monitoring of wind turbines is usually performed using several accelerometers and strain gages attached to the nacelle to monitor the gearbox and equipments. Currently, digital image correlation and stereophotogrammetry are used to measure dynamics of wind turbine blades. These methods usually measure displacement and strain to identify location of defects. Dynamic characteristics of non-rotating wind turbines have been measured using digital image correlation and photogrammetry. Three dimensional point tracking has also been used to measure rotating dynamics of wind turbines.

Materials and Durability

Currently serving wind turbine blades are mainly made of composite materials. These blades are usually made of a polyester resin, a vinyl resin, and epoxy thermosetting matrix resin and E-glass fibers, S- glass fibers and carbon fiber reinforced materials. Construction may use manual layup techniques or composite resin injection molding. As the price of glass fibers is only about one tenth the price of carbon fiber, glass fiber is still dominant. One of the predominant ways wind turbines have gain performance is by increasing rotor diameters, and thus blade length. Longer blades place more demands on the strength and stiffness of the materials. Stiffness is especially important to avoid having blades flex to the degree that they hit the tower of the wind turbine. Carbon fiber is between 4 and 6 times stiffer than glass fiber, so carbon fiber is becoming more common in wind turbine blades.

Wind Turbines on Public Display

A few localities have exploited the attention-getting nature of wind turbines by placing them on public display, either with visitor centers around their bases, or with viewing areas farther away. The wind turbines are generally of conventional horizontal-axis, three-bladed design, and generate power to feed electrical grids, but they also serve the unconventional roles of technology demonstration, public relations, and education.

The Nordex N50 wind turbine and visitor centre of Lamma Winds in Hong Kong, China

Small wind Turbines

Small wind turbines may be used for a variety of applications including on- or off-grid residences, telecom towers, offshore platforms, rural schools and clinics, remote monitoring and other purposes that require energy where there is no electric grid, or where the grid is unstable. Small wind turbines may be as small as a fifty-watt generator for boat or caravan use. Hybrid solar and wind powered units are increasingly being used for traffic signage, particularly in rural locations, as they avoid the need to lay long cables from the nearest mains connection point. The U.S. Department of Energy's National Renewable Energy Laboratory (NREL) defines small wind turbines as those smaller than or equal to 100 kilowatts. Small units often have direct drive generators, direct current output, aeroelastic blades, lifetime bearings and use a vane to point into the wind.

A small Quietrevolution QR5 Gorlov type vertical axis wind turbine in Bristol, England. Measuring 3 m in diameter and 5 m high, it has a nameplate rating of 6.5 kW to the grid.

Larger, more costly turbines generally have geared power trains, alternating current output, flaps and are actively pointed into the wind. Direct drive generators and aeroelastic blades for large wind turbines are being researched.

Wind Turbine Spacing

On most horizontal windturbine farms, a spacing of about 6-10 times the rotor diameter is often upheld. However, for large wind farms distances of about 15 rotor diameters should be more economically optimal, taking into account typical wind turbine and land costs. This conclusion has been reached by research conducted by Charles Meneveau of the Johns Hopkins University, and Johan Meyers of Leuven University in Belgium, based on computer simulations that take into account the detailed interactions among wind turbines (wakes) as well as with the entire turbulent atmospheric boundary layer. Moreover, recent research by John Dabiri of Caltech suggests that vertical wind turbines may be placed much more closely together so long as an alternating pattern of rotation is created allowing blades of neighbouring turbines to move in the same direction as they approach one another.

Operability

Maintenance

Wind turbines need regular maintenance to stay reliable and available, reaching 98%.

Modern turbines usually have a small onboard crane for hoisting maintenance tools and minor components. However, large heavy components like generator, gearbox, blades and so on are rarely replaced and a heavy lift external crane is needed in those cases. If the turbine has a difficult access road, a containerized crane can be lifted up by the internal crane to provide heavier lifting.

Repowering

Installation of new wind turbines can be controversial. An alternative is repowering, where existing wind turbines are replaced with bigger, more powerful ones, sometimes in smaller numbers while keeping or increasing capacity.

Demolition

Older turbines were in some early cases not required to be removed when reaching the end of their life. Some still stand, waiting to be recycled or repowered.

A demolition industry develops to recycle offshore turbines at a cost of DKK 2–4 million per MW, to be guaranteed by the owner.

Records

Largest Capacity Conventional Drive

> The Vestas V164 has a rated capacity of 8 MW, has an overall height of 220 m (722 ft), a diameter of 164 m (538 ft), is for offshore use, and is the world's largest-capacity wind turbine since its introduction in 2014. The conventional drive train consist of a main gearbox and a medium speed PM generator. Prototype installed in 2014 at the National Test Center Denmark nearby Østerild. Series production starts end of 2015.

Largest Capacity Direct Drive

> The Enercon E-126 with 7.58 MW and 127 m rotor diameter is the largest direct drive turbine. It's only for onshore use. The turbine has parted rotor blades with 2 sections for transport. In July 2016, Siemens upgraded its 7 to 8 MW.

Largest Vertical-axis

> Le Nordais wind farm in Cap-Chat, Quebec has a vertical axis wind turbine (VAWT) named Éole, which is the world's largest at 110 m. It has a nameplate capacity of 3.8 MW.

Fuhrländer Wind Turbine Laasow, in Brandenburg, Germany, among the world's tallest wind turbines

Éole, the largest vertical axis wind turbine, in Cap-Chat, Quebec, Canada

Largest 1-bladed Turbine

Riva Calzoni M33 was a single-bladed wind turbine with 350 kW, designed and built In Bologna in 1993.

Largest 2-bladed Turbine

Today's biggest 2-bladed turbine is build by Mingyang Wind Power in 2013. It is a SCD6.5MW offshore downwind turbine, designed by aerodyn Energiesysteme.

Largest Swept Area

The turbine with the largest swept area is the Samsung S7.0-171, with a diameter of 171 m, giving a total sweep of 22966 m².

Tallest

A Nordex 3.3 MW was installed in July 2016. It has a total height of 230m, and a hub height of 164m on 100m concrete tower bottom with steel tubes on top (hybrid tower).

Vestas V164 was the tallest wind turbine, standing in Østerild, Denmark, 220 meters tall, constructed in 2014. It has a steel tube tower.

Highest Tower

Fuhrländer installed a 2.5MW turbine on a 160m lattice tower in 2003 (see Fuhrländer Wind Turbine Laasow and Nowy Tomyśl Wind Turbines).

Most Rotors

Lagerwey has build Four-in-One, a multi rotor wind turbine with one tower and four rotors near Maasvlakte. In April 2016, Vestas installed a 900 kW quadrotor test wind turbine at Risø, made from 4 recycled 225 kW V29 turbines.

Most Productive

Four turbines at Rønland wind farm in Denmark share the record for the most productive wind turbines, with each having generated 63.2 GWh by June 2010.

Highest-situated

Since 2013 the world's highest-situated wind turbine was made and installed by WindAid and is located at the base of the Pastoruri Glacier in Peru at 4,877 meters (16,001 ft) above sea level. The site uses the WindAid 2.5 kW wind generator to supply power to a small rural community of micro entrepreneurs who cater to the tourists who come to the Pastoruri glacier.

Largest Floating Wind Turbine

The world's largest—and also the first operational deep-water large-capacity—floating wind turbine is the 2.3 MW Hywind currently operating 10 kilometers (6.2 mi) offshore in 220-meter-deep water, southwest of Karmøy, Norway. The turbine began operating in September 2009 and utilizes a Siemens 2.3 MW turbine.

Electromagnetic Induction

Electromagnetic or Magnetic induction is the production of an electromotive force or voltage across an electrical conductor due to its dynamic interaction with a magnetic field.

Michael Faraday is generally credited with the discovery of induction in 1831, and James Clerk Maxwell mathematically described it as Faraday's law of induction. Lenz's law describes the direction of the induced field. Faraday's law was later generalized to the Maxwell-Faraday equation, which is one of the equations in James Clerk Maxwell's theory of electromagnetism.

Electromagnetic induction has found many applications in technology, including electrical components such as inductors and transformers, and devices such as electric motors and generators.

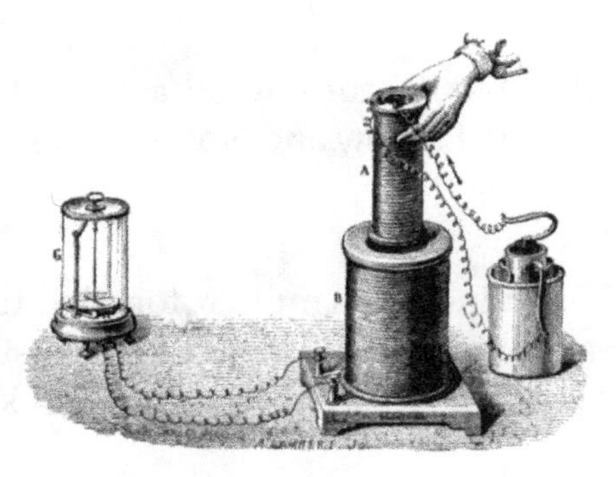

Faraday's experiment showing induction between coils of wire: The liquid battery (right) provides a current that flows through the small coil (A), creating a magnetic field. When the coils are stationary, no current is induced. But when the small coil is moved in or out of the large coil (B), the magnetic flux through the large coil changes, inducing a current which is detected by the galvanometer (G).

History

Electromagnetic induction was first discovered by Michael Faraday, who made his discovery public in 1831. It was discovered independently by Joseph Henry in 1832.

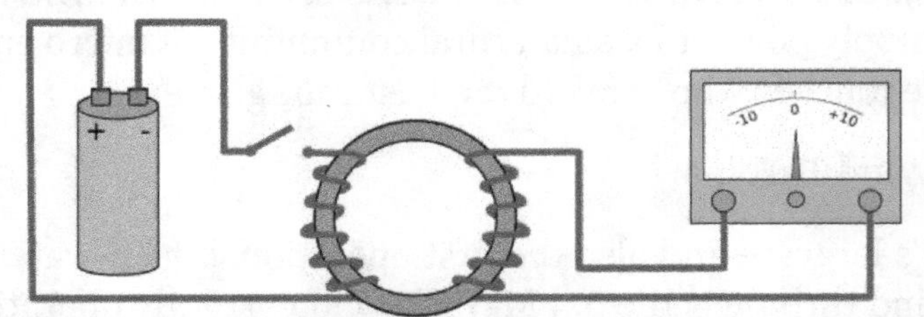

A diagram of Faraday's iron ring apparatus. Change in the magnetic flux of the left coil induces a current in the right coil.

In Faraday's first experimental demonstration (August 29, 1831), he wrapped two wires around opposite sides of an iron ring or "torus" (an arrangement similar to a modern toroidal transformer). Based on his understanding of electromagnets, he expected that, when current started to flow in one wire, a sort of wave would travel through the ring and cause some electrical effect on the opposite side. He plugged one wire into a galvanometer, and watched it as he connected the other wire to a battery. He saw a transient current, which he called a "wave of electricity", when he connected the wire to the battery and another when he disconnected it. This induction was due to the change in magnetic flux that occurred when the battery was connected and disconnected. Within two months, Faraday found several other manifestations of electromagnetic induction. For example, he saw transient currents when he quickly slid a bar magnet in and out of a coil of wires, and he generated a steady (DC) current by rotating a copper disk near the bar magnet with a sliding electrical lead ("Faraday's disk").

Faraday explained electromagnetic induction using a concept he called lines of force. However, scientists at the time widely rejected his theoretical ideas, mainly because they were not formulated mathematically. An exception was James Clerk Maxwell, who used Faraday's ideas as the basis of his quantitative electromagnetic theory. In Maxwell's model, the time varying aspect of electromagnetic induction is expressed as a differential equation, which Oliver Heaviside referred to as Faraday's law even though it is slightly different from Faraday's original formulation and does not describe motional EMF. Heaviside's version is the form recognized today in the group of equations known as Maxwell's equations.

Faraday's disk (see homopolar generator)

In 1834 Heinrich Lenz formulated the law named after him to describe the "flux through the circuit". Lenz's law gives the direction of the induced EMF and current resulting from electromagnetic induction.

Theory

Faraday's Law of Induction and Lenz's Law

Faraday's law of induction makes use of the magnetic flux Φ_B through a region of space enclosed by a wire loop. The magnetic flux is defined by a surface integral:

$$\Phi_B = \int_{\Sigma} \mathbf{B} \cdot d\mathbf{A},$$

where dA is an element of the surface Σ enclosed by the wire loop, B is the magnetic field. The dot product B·dA corresponds to an infinitesimal amount of magnetic flux. In more visual terms, the magnetic flux through the wire loop is proportional to the number of magnetic flux lines that pass through the loop.

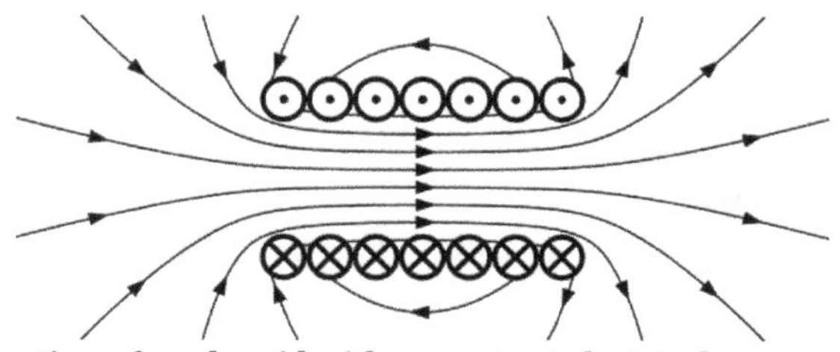

The longitudinal cross section of a solenoid with a constant electrical current running through it. The magnetic field lines are indicated, with their direction shown by arrows. The magnetic flux corresponds to the 'density of field lines'. The magnetic flux is thus densest in the middle of the solenoid, and weakest outside of it.

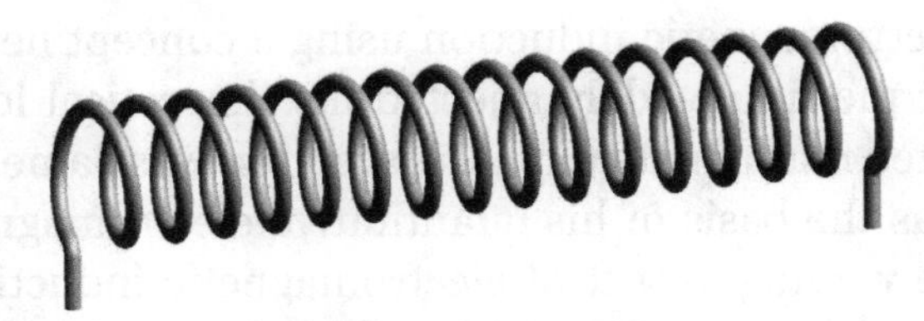

A solenoid

When the flux through the surface changes, Faraday's law of induction says that the wire loop acquires an electromotive force (EMF). The most widespread version of this law states that the induced electromotive force in any closed circuit is equal to the rate of change of the magnetic flux enclosed by the circuit:

$$\mathcal{E} = -\frac{d\Phi_B}{dt},$$

where $\mathcal{E}$ is the EMF and Φ_B is the magnetic flux. The direction of the electromotive force is given by Lenz's law which states that an induced current will flow in the direction that will oppose the change which produced it. This is due to the negative sign in the previous equation. To increase the generated EMF, a common approach is to exploit flux linkage by creating a tightly wound coil of wire, composed of N identical turns, each with the same magnetic flux going through them. The resulting EMF is then N times that of one single wire.

$$\mathcal{E} = -N\frac{d\Phi_B}{dt}$$

Generating an EMF through a variation of the magnetic flux through the surface of a wire loop can be achieved in several ways:

1. the magnetic field B changes (e.g. an alternating magnetic field, or moving a wire loop towards a bar magnet where the B field is stronger),
2. the wire loop is deformed and the surface Σ changes,
3. the orientation of the surface dA changes (e.g. spinning a wire loop into a fixed magnetic field),
4. any combination of the above

Maxwell–Faraday Equation

In general, the relation between the EMF $\mathcal{E}$ in a wire loop encircling a surface Σ, and the electric field E in the wire is given by

$$\mathcal{E} = \oint_{\partial\Sigma} \mathbf{E} \cdot d\ell$$

where $d\ell$ is an element of contour of the surface Σ, combining this with the definition of flux

$$\Phi_B = \int_{\acute{o}} \mathbf{B} \cdot d\mathbf{A},$$

we can write the integral form of the Maxwell–Faraday equation

$$\oint_{\partial\Sigma} \mathbf{E} \cdot d\ell = -\frac{d}{dt}\int_{\Sigma} \mathbf{B} \cdot d\mathbf{A}$$

It is one of the four Maxwell's equations, and therefore plays a fundamental role in the theory of classical electromagnetism.

Faraday's Law and Relativity

Faraday's law describes two different phenomena: the motional EMF generated by a magnetic force on a moving wire , and the transformer EMF generated by an electric force due to a changing magnetic field (due to the differential form of the Maxwell–Faraday equation). James Clerk Maxwell drew attention to the separate physical phenomena in 1861. This is believed to be a unique example in physics of where such a fundamental law is invoked to explain two such different phenomena.

Einstein noticed that the two situations both corresponded to a relative movement between a conductor and a magnet, and the outcome was unaffected by which one was moving. This was one of the principal paths that led him to develop special relativity.

Applications

The principles of electromagnetic induction are applied in many devices and systems, including:

- Current clamp
- Electric generators
- Electromagnetic forming
- Graphics tablet
- Hall effect meters
- Induction cooking
- Induction motors
- Induction sealing
- Induction welding
- Inductive charging

- Inductors
- Magnetic flow meters
- Mechanically powered flashlight
- Pickups
- Rowland ring
- Transcranial magnetic stimulation
- Transformers
- Wireless energy transfer

Electrical Generator

The EMF generated by Faraday's law of induction due to relative movement of a circuit and a magnetic field is the phenomenon underlying electrical generators. When a permanent magnet is moved relative to a conductor, or vice versa, an electromotive force is created. If the wire is connected through an electrical load, current will flow, and thus electrical energy is generated, converting the mechanical energy of motion to electrical energy. For example, the drum generator is based upon the figure to the right. A different implementation of this idea is the Faraday's disc, shown in simplified form on the right.

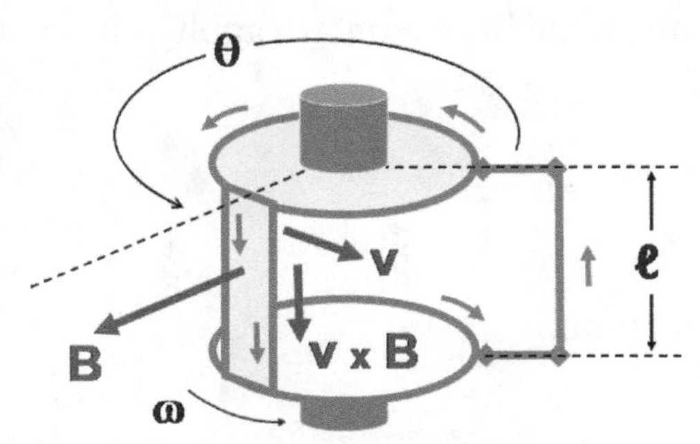

Rectangular wire loop rotating at angular velocity ω in radially outward pointing magnetic field B of fixed magnitude. The circuit is completed by brushes making sliding contact with top and bottom discs, which have conducting rims. This is a simplified version of the drum generator.

In the Faraday's disc example, the disc is rotated in a uniform magnetic field perpendicular to the disc, causing a current to flow in the radial arm due to the Lorentz force. It is interesting to understand how it arises that mechanical work is necessary to drive this current. When the generated current flows through the conducting rim, a magnetic field is generated by this current through Ampère's circuital law (labeled "induced B" in the figure). The rim thus becomes an electromagnet that resists rotation of the disc (an example of Lenz's law). On the far side of the figure, the return current flows from the rotating arm through the far side of the rim to the bottom brush. The B-field induced by this return current opposes the applied B-field, tending to decrease the flux through that side of the circuit, opposing the increase in flux due to rotation. On the near side of the figure, the return current flows from the rotating arm through the near side of the rim to the bottom brush. The induced B-field increases the flux on this side of the circuit, opposing the decrease in flux due to rotation. Thus, both sides of the circuit generate an EMF opposing the rotation. The energy required to keep the disc moving, despite this reactive force, is exactly equal to the electrical energy generated (plus energy wasted due to friction, Joule heating, and other inefficiencies). This behavior is common to all generators converting mechanical energy to electrical energy.

Electrical Transformer

When the electric current in a loop of wire changes, the changing current creates a changing magnetic field. A second wire in reach of this magnetic field will experience this change in magnetic field as a change in its coupled magnetic flux, $d\,\Phi_B / d\,t$. Therefore, an electromotive force is set up in the second loop called the induced EMF or transformer EMF. If the two ends of this loop are connected through an electrical load, current will flow.

Current Clamp

A current clamp is a type of transformer with a split core which can be spread apart and clipped onto a wire or coil to either measure the current in it or, in reverse, to induce a voltage. Unlike conventional instruments the clamp does not make electrical contact with the conductor or require it to be disconnected during attachment of the clamp.

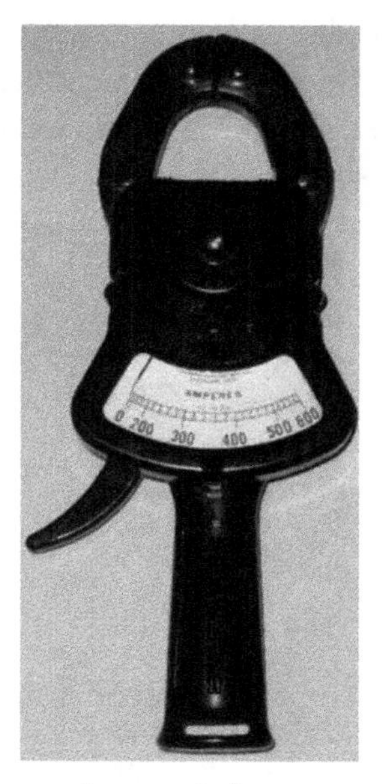

A current clamp

Magnetic Flow Meter

Faraday's law is used for measuring the flow of electrically conductive liquids and slurries. Such instruments are called magnetic flow meters. The induced voltage Ɛ generated in the magnetic field B due to a conductive liquid moving at velocity v is thus given by:

$$\mathcal{E} = -B\ell v,$$

where ℓ is the distance between electrodes in the magnetic flow meter.

Eddy Currents

Conductors (of finite dimensions) moving through a uniform magnetic field, or stationary within a changing magnetic field, will have currents induced within them. These induced eddy currents can be undesirable, since they dissipate energy in the resistance of the conductor. There are a number of methods employed to control these undesirable inductive effects.

- Electromagnets in electric motors, generators, and transformers do not use solid metal, but instead use thin sheets of metal plate, called laminations. These thin plates reduce the parasitic eddy currents, as described below.
- Inductive coils in electronics typically use magnetic cores to minimize parasitic current flow. They are a mixture of metal powder plus a resin binder that can hold any shape. The binder prevents parasitic current flow through the powdered metal.

Electromagnet Laminations

Only five laminations or plates are shown in this example, so as to show the subdivision of the eddy currents. In practical use, the number of laminations or punchings ranges from 40 to 66 per inch, and brings the eddy current loss down to about one percent. While the plates can be separated by insulation, the voltage is so low that the natural rust/oxide coating of the plates is enough to prevent current flow across the laminations.

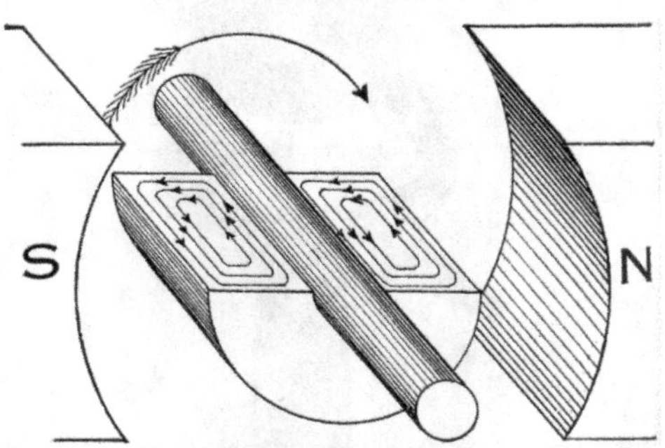

Eddy currents occur when a solid metallic mass is rotated in a magnetic field, because the outer portion of the metal cuts more lines of force than the inner portion, hence the induced electromotive force not being uniform, tends to set up currents between the points of greatest and least potential. Eddy currents consume a considerable amount of energy and often cause a harmful rise in temperature.

This is a rotor approximately 20mm in diameter from a DC motor used in a CD player. Note the laminations of the electromagnet pole pieces, used to limit parasitic inductive losses.

Parasitic induction Within Conductors

High current power-frequency devices, such as electric motors, generators and transformers, use multiple small conductors in parallel to break up the eddy flows that can form within large solid conductors. The same principle is applied to transformers used at higher than power frequency, for example, those used in switch-mode power supplies and the intermediate frequency coupling transformers of radio receivers.

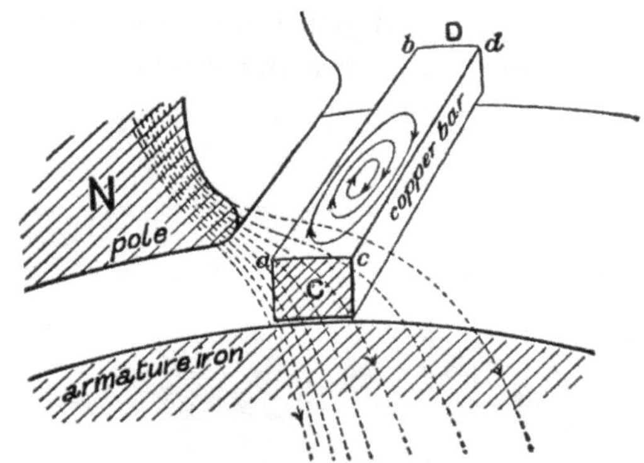

In this illustration, a solid copper bar conductor on a rotating armature is just passing under the tip of the pole piece N of the field magnet. Note the uneven distribution of the lines of force across the copper bar. The magnetic field is more concentrated and thus stronger on the left edge of the copper bar (a,b) while the field is weaker on the right edge (c,d). Since the two edges of the bar move with the same velocity, this difference in field strength across the bar creates whorls or current eddies within the copper bar.

Photovoltaics

Photovoltaics (PV) covers the conversion of light into electricity using semiconducting materials that exhibit the photovoltaic effect, a phenomenon studied in physics, photochemistry, and electrochemistry.

A typical photovoltaic system employs solar panels, each comprising a number of solar cells, which generate electrical power. The first step is the photoelectric effect followed by an electrochemical process where crystallized atoms, ionized in a series, generate an electric current. PV Installations may be ground-mounted, rooftop mounted or wall mounted.

Photovoltaic SUDI shade is an autonomous and mobile station in France that provides energy for electric vehicles using solar energy.

Solar PV generates no pollution. The direct conversion of sunlight to electricity occurs without any moving parts. Photovoltaic systems have been used for fifty years in specialized applications, standalone and grid-connected PV systems have been in use for more than twenty years. They were first mass-produced in 2000, when German environmentalists and the Eurosolar organization got government funding for a ten thousand roof program.

The Solar Settlement, a sustainable housing community project in Freiburg, Germany.

On the other hand, grid-connected PV systems have the major disadvantage that the power output is dependent on direct sunlight, so about 10-25% is lost if a tracking system is not used, since the cell wil not be directly facing the sun at all times.Power output is also adversely affected by weather conditions, especially cloud cover. This means that, in the national grid for example, this power has to be made up by other power sources: hydrocarbon, nuclear, hydroelectric or wind energy. To some, solar installations also have a negative aesthetic impact on an area.

Solar panels on the International Space Station

Advances in technology and increased manufacturing scale have reduced the cost, increased the reliability, and increased the efficiency of photovoltaic instalations and the levelised cost of electricity from PV is competitive, on a kilowatt/ hour basis, with conventional electricity sources in an expanding list of geographic regions. Solar PV regularly costs USD 0.05-0.10 per kilowatt-hour (kWh) in Europe, China, India, South Africa and the United States. In 2015, record low prices were set in the United Arab Emirates (5.84 cents/kWh), Peru (4.8 cents/kWh) and Mexico (4.8 cents/kWh). In May 2016, a solar PV auction in Dubai attracted a bid of 3 cents/kWh.

Double glass photovoltaic solar modules, installed in a support structure.

Net metering and financial incentives, such as preferential feed-in tariffs for solar-generated electricity, have supported solar PV installations in many countries. More than 100 countries now use solar PV. After hydro and wind powers, PV is the third renewable energy source in terms of globally capacity. In 2014, worldwide installed PV capacity increased to 177 gigawatts (GW), which is two percent of global electricity demand. China, followed by Japan and the United States, is the fastest growing market, while Germany remains the world's largest producer (both in per capita and absolute terms), with solar PV providing seven percent of annual domestic electricity consumption.

With current technology (as of 2013), photovoltaics recoups the energy needed to manufacture them in 1.5 years in Southern Europe and 2.5 years in Northern Europe.[T]

Etymology

The term "photovoltaic" comes from the Greek φ□ς (phōs) meaning "light", and from "volt", the unit of electro-motive force, the volt, which in turn comes from the last name of the Italian physicist Alessandro Volta, inventor of the battery (electrochemical cell). The term "photo-voltaic" has been in use in English since 1849.

Solar Cells

Photovoltaics are best known as a method for generating electric power by using solar cells to convert energy from the sun into a flow of electrons. The photovoltaic effect refers to photons of light exciting electrons into a higher state of energy, allowing them to act as charge carriers for an electric current. The photovoltaic effect was first observed by Alexandre-Edmond Becquerel in 1839. The term photovoltaic denotes the unbiased operating mode of a photodiode in which current through the device is entirely due to the transduced light energy. Virtually all photovoltaic devices are some type of photodiode.

Solar cells generate electricity directly from sunlight.

Solar cells produce direct current electricity from sun light which can be used to power equipment or to recharge a battery. The first practical application of photovoltaics was to power orbiting satellites and other spacecraft, but today the majority of photovoltaic modules are used for grid connected power generation. In this case an inverter is required to convert the DC to AC. There is a smaller market for off-grid power for remote dwellings, boats, recreational vehicles, electric cars, roadside emergency telephones, remote sensing, and cathodic protection of pipelines.

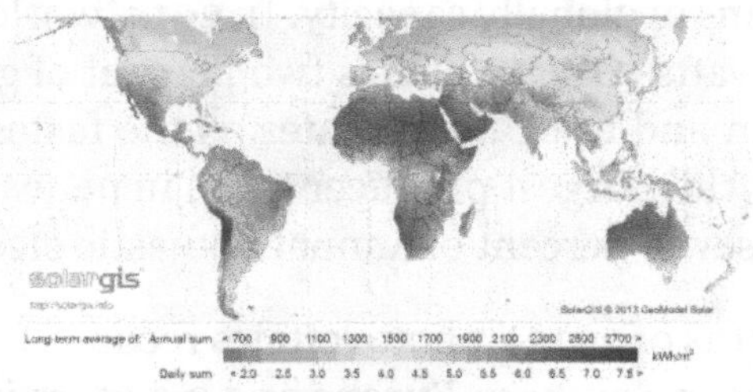

Average insolation. Note that this is for a horizontal surface. Solar panels are normally propped up at an angle and receive more energy per unit area.

Photovoltaic power generation employs solar panels composed of a number of solar cells containing a photovoltaic material. Materials presently used for photovoltaics include monocrystalline silicon, polycrystalline silicon, amorphous silicon, cadmium telluride, and copper indium gallium selenide/sulfide. Copper solar cables connect modules (module cable), arrays (array cable), and sub-fields. Because of the growing demand for renewable energy sources, the manufacturing of solar cells and photovoltaic arrays has advanced considerably in recent years.

Solar photovoltaics power generation has long been seen as a clean energy technology which draws upon the planet's most plentiful and widely distributed renewable energy source – the sun. The technology is "inherently elegant" in that the direct conversion of sunlight to electricity occurs without any moving parts or environmental emissions during operation. It is well proven, as photovoltaic systems have now been used for fifty years in specialised applications, and grid-connected systems have been in use for over twenty years.

Cells require protection from the environment and are usually packaged tightly behind a glass sheet. When more power is required than a single cell can deliver, cells are electrically connected together to form photovoltaic modules, or solar panels. A single module is enough to power an emergency telephone, but for a house or a power plant the modules must be arranged in multiples as arrays.

Photovoltaic power capacity is measured as maximum power output under standardized test conditions (STC) in "W_p" (watts peak). The actual power output at a particular point in time may be less than or greater than this standardized, or "rated," value, depending on geographical location, time of day, weather conditions, and other factors. Solar photovoltaic array capacity factors are typically under 25%, which is lower than many other industrial sources of electricity.

Current Developments

For best performance, terrestrial PV systems aim to maximize the time they face the sun. Solar trackers achieve this by moving PV panels to follow the sun. The increase can be by as much as 20% in winter and by as much as 50% in summer. Static mounted systems can be optimized by analysis of the sun path. Panels are often set to latitude tilt, an angle equal to the latitude, but performance can be improved by adjusting the angle for summer or winter. Generally, as with other semiconductor devices, temperatures above room temperature reduce the performance of photovoltaics.

A number of solar panels may also be mounted vertically above each other in a tower, if the zenith distance of the Sun is greater than zero, and the tower can be turned horizontally as a whole and each panels additionally around a horizontal axis. In such a tower the panels can follow the Sun exactly. Such a device may be described as a ladder mounted on a turnable disk. Each step of that ladder is the middle axis of a rectangular solar panel. In case the zenith distance of the Sun reaches zero, the "ladder" may be rotated to the north or the south to avoid a solar panel producing a shadow on a lower solar panel. Instead of an exactly vertical tower one can choose a tower with an axis directed to the polar star, meaning that it is parallel to the rotation axis of the Earth. In this case the angle between the axis and the Sun is always larger than 66 degrees. During a day it is only necessary to turn the panels around this axis to follow the Sun. Installations may be ground-mounted (and sometimes integrated with farming and grazing) or built into the roof or walls of a building (building-integrated photovoltaics).

Another recent development involves the makeup of solar cells. Perovskite is a very inexpensive material which is being used to replace the expensive crystalline silicon which is still part of a standard PV cell build to this day. Michael Graetzel, Director of the Laboratory of Photonics and Interfaces at EPFL says, “Today, efficiency has peaked at 18 percent, but it's expected to get even higher in the future.” This is a significant claim, as 20% efficiency is typical among solar panels which use more expensive materials.

Efficiency

Electrical efficiency (also called conversion efficiency) is a contributing factor in the selection of a photovoltaic system. However, the most efficient solar panels are typically the most expensive, and may not be commercially available. Therefore, selection is also driven by cost efficiency and other factors.

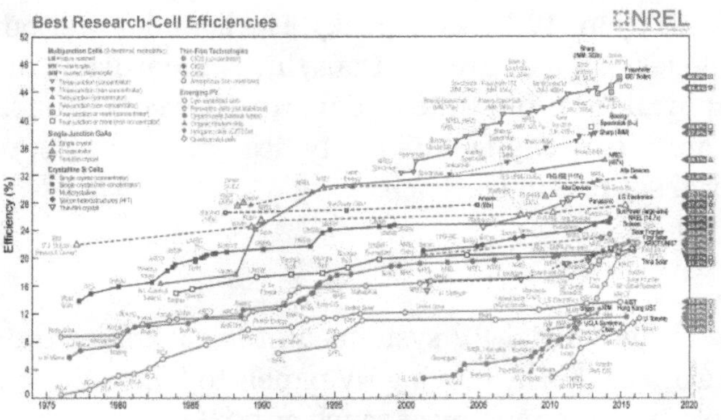

Best Research-Cell Efficiencies

The electrical efficiency of a PV cell is a physical property which represents how much electrical power a cell can produce for a given insolation. The basic expression for maximum efficiency of a photovoltaic cell is given by the ratio of output power to the incident solar power (radiation flux times area)

$$\eta = \frac{P_{max}}{E \cdot A_{cell}}.[26]$$

The efficiency is measured under ideal laboratory conditions and represents the maximum achievable efficiency of the PV material. Actual efficiency is influenced by the output Voltage, current, junction temperature, light intensity and spectrum.

The most efficient type of solar cell to date is a multi-junction concentrator solar cell with an efficiency of 46.0% produced by Fraunhofer ISE in December 2014. The highest efficiencies achieved without concentration include a material by Sharp Corporation at 35.8% using a proprietary triple-junction manufacturing technology in 2009, and Boeing Spectrolab (40.7% also using a triple-layer design). The US company SunPower produces cells that have an efficiency of 21.5%, well above the market average of 12–18%.

There is an ongoing effort to increase the conversion efficiency of PV cells and modules, primarily for competitive advantage. In order to increase the efficiency of solar cells, it is important to choose a semiconductor material with an appropriate band gap that matches the solar spectrum. This will enhance the electrical and optical properties. Improving the method of charge collection is also useful for increasing the efficiency. There are several groups of materials that are being developed. Ultrahigh-efficiency devices ($\eta > 30\%$) are made by using GaAs and GaInP2 semiconductors with multijunction tandem cells. High-quality, single-crystal silicon materials are used to achieve high-efficiency, low cost cells ($\eta > 20\%$).

Recent developments in Organic photovoltaic cells (OPVs) have made significant advancements in power conversion efficiency from 3% to over 15% since their introduction in the 1980s. To date, the highest reported power conversion efficiency ranges from 6.7% to 8.94% for small molecule, 8.4%–10.6% for polymer OPVs, and 7% to 21% for perovskite OPVs. OPV's are expected to play a major role in the PV market. Recent improvements have increased the efficiency and lowered cost, while remaining environmentally-benign and renewable.

Several companies have begun embedding power optimizers into PV modules called smart modules. These modules perform maximum power point tracking (MPPT) for each module individually, measure performance data for monitoring, and provide additional safety features. Such modules can also compensate for shading effects, wherein a shadow falling across a section of a module causes the electrical output of one or more strings of cells in the module to decrease.

One of the major causes for the decreased performance of cells is overheating. The efficiency of a solar cell declines by about 0.5% for every 1 degree Celsius increase in temperature. This means that a 100 degree increase in surface temperature could decrease the efficiency of a solar cell by about half. Self-cooling solar cells are one solution to this problem. Rather than using energy to cool the surface, pyramid and cone shapes can be formed from silica, and attached to the surface of a solar panel. Doing so allows visible light to reach the solar cells, but reflects infrared rays (which carry heat).

Growth

Solar photovoltaics is growing rapidly and worldwide installed capacity reached at least 177 gigawatts (GW) by the end of 2014. The total power output of the world's PV capacity in a calendar year is now beyond 200 TWh of electricity. This represents 1% of worldwide electricity demand. More than 100 countries use solar PV. China, followed by Japan and the United States is now the fastest growing market, while Germany remains the world's largest producer, contributing more than 7% to its national electricity demands. Photovoltaics is now, after hydro and wind power, the third most important renewable energy source in terms of globally installed capacity.

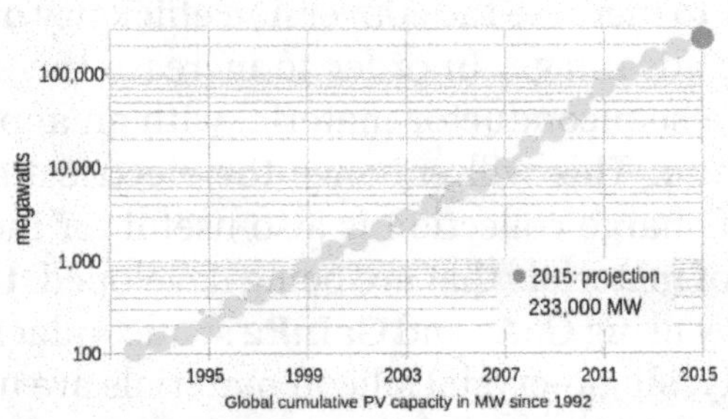

Worldwide growth of photovoltaics on a semi-log plot since 1992

Projected Global Growth (MW)

Several market research and financial companies foresee record-breaking global installation of more than 50 GW in 2015. China is predicted to take the lead from Germany and to become the world's largest producer of PV power by installing another targeted 17.8 GW in 2015. India is expected to install 1.8 GW, doubling its annual installations. By 2018, worldwide photovoltaic capacity is projected to doubled or even triple to 430 GW. Solar Power Europe (formerly known as EPIA) also estimates that photovoltaics will meet 10% to 15% of Europe's energy demand in 2030.

The EPIA/Greenpeace Solar Generation Paradigm Shift Scenario (formerly called Advanced Scenario) from 2010 shows that by the year 2030, 1,845 GW of PV systems could be generating approximately 2,646 TWh/year of electricity around the world. Combined with energy use efficiency improvements, this would represent the electricity needs of more than 9% of the world's population. By 2050, over 20% of all electricity could be provided by photovoltaics.

Michael Liebreich, from Bloomberg New Energy Finance, anticipates a tipping point for solar energy. The costs of power from wind and solar are already below those of conventional electricity generation in some parts of the world, as they have fallen sharply and will continue to do so. He also asserts, that the electrical grid has been greatly expanded worldwide, and is ready to receive and distribute electricity from renewable sources. In addition, worldwide electricity prices came under strong pressure from renewable energy sources, that are, in part, enthusiastically embraced by consumers.

Deutsche Bank sees a "second gold rush" for the photovoltaic industry to come. Grid parity has already been reached in at least 19 markets by January 2014. Photovoltaics will prevail beyond feed-in tariffs, becoming more competitive as deployment increases and prices continue to fall.

In June 2014 Barclays downgraded bonds of U.S. utility companies. Barclays expects more competition by a growing self-consumption due to a combination of decentralized PV-systems and residential electricity storage. This could fundamentally change the utility's business model and transform the system over the next ten years, as prices for these systems are predicted to fall.

Environmental Impacts of Photovoltaic Technologies

Types of Impacts

While solar photovoltaic (PV) cells are promising for clean energy production, their deployment is hindered by production costs, material availability, and toxicity. Life cycle assessment (LCA) is one method of determining environmental impacts from PV. Many studies have been done on the various types of PV including first generation, second generation, and third generation. Usually these PV LCA studies select a cradle to gate system boundary because often at the time the studies are conducted, it is a new technology not commercially available yet and their required balance of system components and disposal methods are unknown.

A traditional LCA can look at many different impact categories ranging from global warming potential, eco-toxicity, human toxicity, water depletion, and many others. Most LCAs of PV have focused on two categories: carbon dioxide equivalents per kWh and energy pay-back time (EPBT). The EPBT is defined as " the time needed to compensate for the total renewable- and non-renewable- primary energy required during the life cycle of a PV system". A 2015 review of EPBT from first and second generation PV suggested that there was greater variation in embedded energy than in efficiency of the cells implying that it was mainly the embedded energy that needs to reduce to have a greater reduction in EPBT. One difficulty in determining impacts due to PV is to determine if the wastes are released to the air, water, or soil during the manufacturing phase. Research is underway to try to understand emissions and releases during the lifetime of PV systems.

Impacts from First-generation PV

Crystalline silicon modules are the most extensively studied PV type in terms of LCA since they are the most commonly used. Mono-crystalline silicon photovoltaic systems (mono-si) have an average efficiency of 14.0%. The cells tend to follow a structure of front electrode, anti-reflection film, n-layer, p-layer, and back electrode, with the sun hitting the front electrode. EPBT ranges from 1.7 to 2.7 years. The cradle to gate of CO_2-eq/kWh ranges from 37.3 to 72.2 grams.

Techniques to produce multi-crystalline silicon (multi-si) photovoltaic cells are simpler and cheaper than mono-si, however tend to make less efficient cells, an average of 13.2%. EPBT ranges from 1.5 to 2.6 years. The cradle to gate of CO_2-eq/kWh ranges from 28.5 to 69 grams. Some studies have looked beyond EPBT and GWP to other environmental impacts. In one such study, conventional energy mix in Greece was compared to multi-si PV and found a 95% overall reduction in impacts including carcinogens, eco-toxicity, acidification, eutrophication, and eleven others.

Impacts from Second Generation

Cadmium telluride (CdTe) is one of the fastest-growing thin film based solar cells which are collectively known as second generation devices. This new thin film device also shares similar performance restrictions (Shockley-Queisser efficiency limit) as conventional Si devices but promises to lower the cost of each device by both reducing material and energy consumption during manufacturing. Today the global market share of CdTe is 5.4%, up from 4.7% in 2008. This technology's highest power conversion efficiency is 21%. The cell structure includes glass substrate (around 2 mm), transparent conductor layer, CdS buffer layer (50–150 nm), CdTe absorber and a metal contact layer.

CdTe PV systems require less energy input in their production than other commercial PV systems per unit electricity production. The average CO_2-eq/kWh is around 18 grams (cradle to gate). CdTe has the fastest EPBT of all commercial PV technologies, which varies between 0.3 and 1.2 years.

Copper Indium Gallium Diselenide (CIGS) is a thin film solar cell based on the copper indium diselenide (CIS) family of chalcopyrite semiconductors. CIS and CIGS are often used interchangeably within the CIS/CIGS community. The cell structure includes soda lime glass as the substrate, Mo layer as the back contact, CIS/CIGS as the absorber layer, cadmium sulfide (CdS) or Zn (S,OH)x as the buffer layer, and ZnO:Al as the front contact. CIGS is approximately 1/100th the thickness of conventional silicon solar cell technologies. Materials necessary for assembly are readily available, and are less costly per watt of solar cell. CIGS based solar devices resist performance degradation over time and are highly stable in the field.

Reported global warming potential impacts of CIGS range from 20.5 – 58.8 grams CO_2-eq/kWh of electricity generated for different solar irradiation (1,700 to 2,200 kWh/m^2/y) and power conversion efficiency (7.8 – 9.12%). EPBT ranges from 0.2 to 1.4 years, while harmonized value of EPBT was found 1.393 years. Toxicity is an issue within the buffer layer of CIGS modules because it contains cadmium and gallium. CIS modules do not contain any heavy metals.

Impacts from Third Generation

Third-generation PVs are designed to combine the advantages of both the first and second generation devices and they do not have Shockley-Queisser efficiency limit, a theoretical limit for first and second generation PV cells. The thickness of a third generation device is less than 1 μm.

One emerging alternative and promising technology is based on an organic-inorganic hybrid solar cell made of methylammonium lead halide perovskites. Perovskite PV cells have progressed rapidly over the past few years and have become one of the most attractive areas for PV research. The cell structure includes a metal back contact (which can be made of Al, Au or Ag), a hole transfer layer (spiro-MeOTAD, P3HT, PTAA, CuSCN, CuI, or NiO), and absorber layer ($CH_3NH_3PbIxBr_3$-x, $CH_3NH_3PbIxCl_3$-x or $CH_3NH_3PbI_3$), an electron transport layer (TiO, ZnO, Al_2O_3 or SnO_2) and a top contact layer (fluorine doped tin oxide or tin doped indium oxide).

There are a limited number of published studies to address the environmental impacts of perovskite solar cells. The major environmental concern is the lead used in the ab sorber layer. Due to the instability of perovskite cells lead may eventually be exposed to fresh water during the use phase. These LCA studies looked at human and ecotoxicity of perovskite solar cells and found they were surprisingly low and may not be an envi ronmental issue. Global warming potential of perovskite PVs were found to be in the range of 24–1500 grams CO_2-eq/kWh electricity production. Similarly, reported EPBT of the published paper range from 0.2 to 15 years. The large range of reported values highlight the uncertainties associated with these studies.

Two new promising thin film technologies are copper zinc tin sulfide (Cu_2ZnSnS_4 or CZTS) and zinc phosphide (Zn_3P_2). Both of these thin films are currently only produced in the lab but may be commercialized in the future. Their manufacturing processes are expected to be similar to those of current thin film technologies of CIGS and CdTe, respectively. Yet, contrary to CIGS and CdTe, CZTS and Zn_3P_2 are made from earth abundant, nontoxic materials and have the potential to produce more electricity annually than the current worldwide consumption. While CZTS and Zn_3P_2 offer good promise for these reasons, the specific environmental implications of their commercial production are not yet known. Global warming potential of CZTS and Zn_3P_2 were found 38 and 30 grams CO_2-eq/kWh while their corresponding EPBT were found 1.85 and 0.78 years, respectively. Overall, CdTe and Zn_3P_2 have similar environmental impacts but can slightly outperform CIGS and CZTS.

Organic and polymer photovoltaic (OPV) are a relatively new area of research. The tradition OPV cell structure layers consist of a semi-transparent electrode, electron blocking layer, tunnel junction, holes blocking layer, electrode, with the sun hitting the transparent electrode. OPV replaces silver with carbon as an electrode material lowering manufacturing cost and making them more environmentally friendly. OPV are flexible, low weight, and work well with roll-to roll manufacturing for mass production. OPV uses "only abundant elements coupled to an extremely low embodied energy through very low processing temperatures using only ambient processing conditions on simple printing equipment enabling energy pay-back times". Current efficiencies range from 1–6.5%, however theoretical analyses show promise beyond 10% efficiency.

Many different configurations of OPV exist using different materials for each layer. OPV technology rivals existing PV technologies in terms of EPBT even if they currently present a shorter operational lifetime. A 2013 study analyzed 12 different configurations all with 2% efficiency, the EPBT ranged from 0.29–0.52 years for 1 m^2 of PV. The average CO_2-eq/kWh for OPV is 54.922 grams.

Economics

There have been major changes in the underlying costs, industry structure and market prices of solar photovoltaics technology, over the years, and gaining a coherent picture of the shifts occurring across the industry value chain globally is a challenge. This is due to: "the rapidity of cost and price changes, the complexity of the PV supply chain, which involves a large number of manufacturing processes, the balance of system (BOS) and installation costs associated with complete PV systems, the choice of different distribution channels, and differences between regional markets within which PV is being deployed". Further complexities result from the many different policy support initiatives that have been put in place to facilitate photovoltaics commercialisation in various countries.

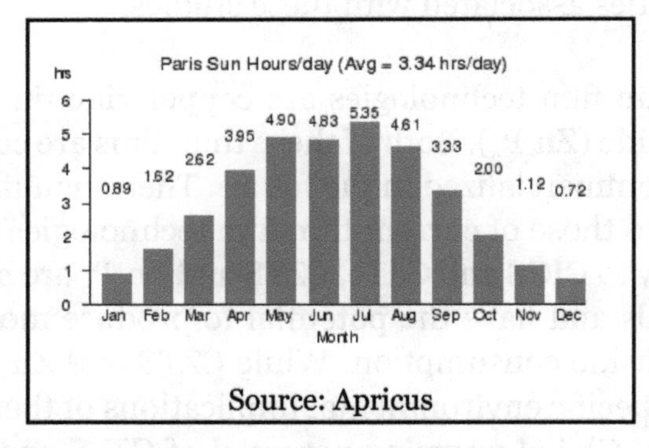

Source: Apricus

The PV industry has seen dramatic drops in module prices since 2008. In late 2011, factory-gate prices for crystalline-silicon photovoltaic modules dropped below the $1.00/W mark. The $1.00/W installed cost, is often regarded in the PV industry as marking the achievement of grid parity for PV. Technological advancements, manufacturing process improvements, and industry re-structuring, mean that further price reductions are likely in coming years.

Financial incentives for photovoltaics, such as feed-in tariffs, have often been offered to electricity consumers to install and operate solar-electric generating systems. Government has sometimes also offered incentives in order to encourage the PV industry to achieve the economies of scale needed to compete where the cost of PV-generated electricity is above the cost from the existing grid. Such policies are implemented to promote national or territorial energy independence, high tech job creation and reduction of carbon dioxide emissions which cause global warming. Due to economies of scale solar panels get less costly as people use and buy more—

as manufacturers increase production to meet demand, the cost and price is expected to drop in the years to come.

Solar cell efficiencies vary from 6% for amorphous silicon-based solar cells to 44.0% with multiple-junction concentrated photovoltaics. Solar cell energy conversion efficiencies for commercially available photovoltaics are around 14–22%. Concentrated photovoltaics (CPV) may reduce cost by concentrating up to 1,000 suns (through magnifying lens) onto a smaller sized photovoltaic cell. However, such concentrated solar power requires sophisticated heat sink designs, otherwise the photovoltaic cell overheats, which reduces its efficiency and life. To further exacerbate the concentrated cooling design, the heat sink must be passive, otherwise the power required for active cooling would reduce the overall efficiency and economy.

Crystalline silicon solar cell prices have fallen from $76.67/Watt in 1977 to an estimated $0.74/Watt in 2013. This is seen as evidence supporting Swanson's law, an observation similar to the famous Moore's Law that states that solar cell prices fall 20% for every doubling of industry capacity.

As of 2011, the price of PV modules has fallen by 60% since the summer of 2008, according to Bloomberg New Energy Finance estimates, putting solar power for the first time on a competitive footing with the retail price of electricity in a number of sunny countries; an alternative and consistent price decline figure of 75% from 2007 to 2012 has also been published, though it is unclear whether these figures are specific to the United States or generally global. The levelised cost of electricity (LCOE) from PV is competitive with conventional electricity sources in an expanding list of geographic regions, particularly when the time of generation is included, as electricity is worth more during the day than at night. There has been fierce competition in the supply chain, and further improvements in the levelised cost of energy for solar lie ahead, posing a growing threat to the dominance of fossil fuel generation sources in the next few years. As time progresses, renewable energy technologies generally get cheaper, while fossil fuels generally get more expensive:

The less solar power costs, the more favorably it compares to conventional power, and the more attractive it becomes to utilities and energy users around the globe. Utility-scale solar power can now be delivered in California at prices well below $100/MWh ($0.10/kWh) less than most other peak generators, even those running on low-cost natural gas. Lower solar module costs also stimulate demand from consumer markets where the cost of solar compares very favorably to retail electric rates.

As of 2011, the cost of PV has fallen well below that of nuclear power and is set to fall further. The average retail price of solar cells as monitored by the Solarbuzz group fell from $3.50/watt to $2.43/watt over the course of 2011.

For large-scale installations, prices below $1.00/watt were achieved. A module price of 0.60 Euro/watt ($0.78/watt) was published for a large scale 5-year deal in April 2012.

By the end of 2012, the "best in class" module price had dropped to $0.50/watt, and was expected to drop to $0.36/watt by 2017.

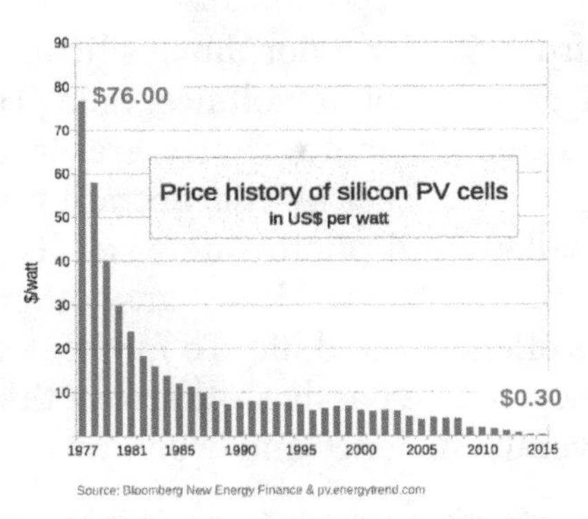

Price per watt history for conventional (c-Si) solar cells since 1977.

In many locations, PV has reached grid parity, which is usually defined as PV production costs at or below retail electricity prices (though often still above the power station prices for coal or gas-fired generation without their distribution and other costs). However, in many countries there is still a need for more access to capital to develop PV projects. To solve this problem securitization has been proposed and used to accelerate development of solar photovoltaic projects. For example, SolarCity offered, the first U.S. asset-backed security in the solar industry in 2013.

Photovoltaic power is also generated during a time of day that is close to peak demand (precedes it) in electricity systems with high use of air conditioning. More generally, it is now evident that, given a carbon price of $50/ton, which would raise the price of coal-fired power by 5c/kWh, solar PV will be cost-competitive in most locations. The declining price of PV has been reflected in rapidly growing installations, totaling about 23 GW in 2011. Although some consolidation is likely in 2012, due to support cuts in the large markets of Germany and Italy, strong growth seems likely to continue for the rest of the decade. Already, by one estimate, total investment in renewables for 2011 exceeded investment in carbon-based electricity generation.

In the case of self consumption payback time is calculated based on how much electricity is not brought from the grid. Additionally, using PV solar power to charge DC batteries, as used in Plug-in Hybrid Electric Vehicles and Electric Vehicles, leads to greater efficiencies. Traditionally, DC generated electricity from solar PV must be converted to AC for buildings, at an average 10% loss during the conversion. An additional efficiency loss occurs in the transition back to DC for battery driven devices and vehicles, and using various interest rates and energy price changes were calculated to find present values that range from $2,057.13 to $8,213.64 (analysis from 2009).

For example, in Germany with electricity prices of 0.25 euro/kWh and Insolation of 900 kWh/kW one kW_p will save 225 euro per year and with installation cost of 1700 euro/kW_p means that the system will pay back in less than 7 years.

Manufacturing

Overall the manufacturing process of creating solar photovoltaics is simple in that it does not require the culmination of many complex or moving parts. Because of the solid state nature of PV systems they often have relatively long lifetimes, anywhere from 10 to 30 years. In order to increase electrical output of a PV system the manufacturer must simply add more photovoltaic components and because of this economies of scale are important for manufacturers as costs decrease with increasing output.

While there are many types of PV systems known to be effective, crystalline silicon PV accounted for around 90% of the worldwide production of PV in 2013. Manufacturing silicon PV systems has several steps. First, polysilicon is processed from mined quartz until it is very pure (semi-conductor grade). This is melted down when small amounts of Boron, a group III element, are added to make a p-type semiconductor rich in electron holes. Typically using a seed crystal, an ingot of this solution is grown from the liquid polycrystalline. The ingot may also be cast in a mold. Wafers of this semiconductor material are cut from the bulk material with wire saws, and then go through surface etching before being cleaned. Next, the wafers are placed into a phosphorus vapor deposition furnace which lays a very thin layer of phosphorus, a group V element, which creates an N-type semiconducting surface. To reduce energy losses an anti-reflective coating is added to the surface, along with electrical contacts. After finishing the cell, cells are connected via electrical circuit according to the specific application and prepared for shipping and installation.

Crystalline silicon photovoltaics are only one type of PV, and while they represent the majority of solar cells produced currently there are many new and promising technologies that have the potential to be scaled up to meet future energy needs.

Another newer technology, thin-film PV, are manufactured by depositing semiconducting layers on substrate in vacuum. The substrate is often glass or stainless-steel, and these semiconducting layers are made of many types of materials including cadmium telluride (CdTe), copper indium diselenide (CIS), copper indium gallium diselenide (CIGS), and amorphous silicon (a-Si). After being deposited onto the substrate the semiconducting layers are separated and connected by electrical circuit by laser-scribing. Thin-film photovoltaics now make up around 20% of the overall production of PV because of the reduced materials requirements and cost to manufacture modules consisting of thin-films as compared to silicon-based wafers.

Other emerging PV technologies include organic, dye-sensitized, quantum-dot, and Perovskite photovoltaics. OPVs fall into the thin-film category of manufacturing, and typically operate around the 12% efficiency range which is lower than the 12–21% typically seen by silicon based PVs. Because organic photovoltaics require very high purity and are relatively reactive they must be encapsulated which vastly increases cost of manufacturing and meaning that they are not feasible for large scale up. Dye-sensitized PVs are similar in efficiency to OPVs but are significantly easier to manufacture. However these dye-sensitized photovoltaics present storage problems because the liquid electrolyte is toxic and can potentially permeate the plastics used in the cell. Quantum dot solar cells are quantum dot sensitized DSSCs and are solution processed meaning they are potentially scalable, but currently they have not reached greater than 10% efficiency. Perovskite solar cells are a very efficient solar energy converter and have excellent optoelectric properties for photovoltaic purposes, but they are expensive and difficult to manufacture.

Applications

Photovoltaic Systems

A photovoltaic system, or solar PV system is a power system designed to supply usable solar power by means of photovoltaics. It consists of an arrangement of several components, including solar panels to absorb and directly convert sunlight into electricity, a solar inverter to change the electric current from DC to AC, as well as mounting, cabling and other electrical accessories. PV systems range from small, roof-top mounted or building-integrated systems with capacities from a few to several tens of kilowatts, to large utility-scale power stations of hundreds of megawatts. Nowadays, most PV systems are grid-connected, while stand-alone systems only account for a small portion of the market.

- Rooftop and building integrated systems

  Photovoltaic arrays are often associated with buildings: either integrated into them, mounted on them or mounted nearby on the ground. Rooftop PV systems are most often retrofitted into existing buildings, usually mounted on top of the existing roof structure or on the existing walls. Alternatively, an array can be located separately from the building but connected by cable to supply power for the building. Building-integrated photovoltaics (BIPV) are increasingly incorporated into the roof or walls of new domestic and industrial buildings as a principal or ancillary source of electrical power. Roof tiles with integrated PV cells are sometimes used as well. Provided there is an open gap in which air can circulate, rooftop mounted solar panels can provide a passive cooling effect on buildings during the day and also keep accumulated heat in at night. Typically, residential rooftop systems have small capacities of around 5–10 kW, while commercial rooftop systems often amount to several hundreds of kilowatts. Although rooftop systems are much smaller than ground-mounted utility-scale power plants, they account for most of the worldwide installed capacity.

Rooftop PV on half-timbered house

- Concentrator photovoltaics

  Concentrator photovoltaics (CPV) is a photovoltaic technology that contrary to conventional flat-plate PV systems uses lenses and curved mirrors to focus sunlight onto small, but highly efficient, multi-junction (MJ) solar cells. In addition, CPV systems often use solar trackers and sometimes a cooling system to further increase their efficiency. Ongoing research and development is rapidly improving their competitiveness in the utility-scale segment and in areas of high solar insolation.

- Photovoltaic thermal hybrid solar collector

  Photovoltaic thermal hybrid solar collector (PVT) are systems that convert solar radiation into thermal and electrical energy. These systems combine a solar PV cell, which converts sunlight into electricity, with a solar thermal collector, which captures the remaining energy and removes waste heat from the PV module. The capture of both electricity and heat allow these devices to have higher exergy and thus be more overall energy efficient than solar PV or solar thermal alone.

- Power stations

  Many utility-scale solar farms have been constructed all over the world. As of 2015, the 579-megawatt (MW_{AC}) Solar Star is the world's largest photovoltaic power station, followed by the Desert Sunlight Solar Farm and the Topaz Solar Farm, both with a capacity of 550 MW_{AC}, constructed by US-company First Solar, using CdTe modules, a thin-film PV technology. All three power stations are located in the Californian desert. Many solar farms around the world are integrated with agriculture and some use innovative solar tracking systems that follow the sun's daily path across the sky to generate more electricity than conventional fixed-mounted systems. There are no fuel costs or emissions during operation of the power stations.

Satellite image of the Topaz Solar Farm

- Rural electrification

  Developing countries where many villages are often more than five kilometers away from grid power are increasingly using photovoltaics. In remote locations in India a rural lighting program has been providing solar powered LED lighting to replace kerosene lamps. The solar powered lamps were sold at about the cost of a few months' supply of kerosene. Cuba is working to provide solar power for areas that are off grid. More complex applications of off-grid solar energy use include 3D printers. RepRap 3D printers have been solar powered with photovoltaic technology, which enables distributed manufacturing for sustainable development. These are areas where the social costs and benefits offer an excellent case for going solar, though the lack of profitability has relegated such endeavors to humanitarian efforts. However, in 1995 solar rural electrification projects had been found to be difficult to sustain due to unfavorable economics, lack of technical support, and a legacy of ulterior motives of north-to-south technology transfer.

- Standalone systems

  Until a decade or so ago, PV was used frequently to power calculators and novelty devices. Improvements in integrated circuits and low power liquid crystal displays make it possible to power such devices for several years between battery changes, making PV use less common. In contrast, solar powered remote fixed devices have seen increasing use recently in locations where significant connection cost makes grid power prohibitively expensive. Such applications include solar lamps, water pumps, parking meters, emergency telephones, trash compactors, temporary traffic signs, charging stations, and remote guard posts and signals.

- Floatovoltaics

  In May 2008, the Far Niente Winery in Oakville, CA pioneered the world's first "floatovoltaic" system by installing 994 photovoltaic solar panels onto 130 pontoons and floating them on the winery's irrigation pond. The floating system generates about 477 kW of peak output and when combined with an array of cells located adjacent to the pond is able to fully offset the winery's electricity consumption. The primary benefit of a floatovoltaic system is that it avoids the need to sacrifice valuable land area that could be used for another purpose. In the case of the Far Niente Winery, the floating system saved three-quarters of an acre that would have been required for a land-based system. That land area can instead be used for agriculture. Another benefit of a floatovoltaic system is that the panels are kept at a lower temperature than they would be on land, leading to a higher efficiency of solar energy conversion. The floating panels also reduce the amount of water lost through evaporation and inhibit the growth of algae.

- In transport

Solar Impulse 2, a solar aircraft

Standalone PV system at an ecotourism resort (British Columbia, Canada).

PV has traditionally been used for electric power in space. PV is rarely used to provide motive power in transport applications, but is being used increasingly to provide auxiliary power in boats and cars. Some automobiles are fitted with solar-powered air conditioning to limit interior temperatures on hot days. A self-contained solar vehicle would have limited power and utility, but a solar-charged electric vehicle allows use of solar power for transportation. Solar-powered cars, boats and airplanes have been demonstrated, with the most practical and likely of these being solar cars. The Swiss solar aircraft, Solar Impulse 2, achieved the longest non-stop solo flight in history and plan to make the first solar-powered aerial circumnavigation of the globe in 2015.

- Telecommunication and Signaling

Solar PV power is ideally suited for telecommunication applications such as local telephone exchange, radio and TV broadcasting, microwave and other forms of electronic communication links. This is because, in most telecommunication application, storage batteries are already in use and the electrical system is basically DC. In hilly and mountainous terrain, radio and TV signals may not reach as they get blocked or reflected back due to undulating terrain. At these locations, low power transmitters (LPT) are installed to receive and retransmit the signal for local population.

- Spacecraft Applications

Part of Juno's solar array

Solar panels on spacecraft are usually the sole source of power to run the sensors, active heating and cooling, and communications. A battery stores this energy for use when the solar panels are in shadow. In some, the power is also used for spacecraft propulsion—electric propulsion. Spacecraft were one of the earliest applications of photovoltaics, starting with the silicon solar cells used on the Vanguard 1 satellite, launched by the US in 1958. Since then, solar power has been used on missions ranging from the MESSENGER probe to Mercury, to as far out in the solar system as the Juno probe to Jupiter. The largest solar power system flown in space is the electrical system of the International Space Station. To increase the power generated per kilogram, typical spacecraft solar panels use high-cost, high-efficiency, and close-packed rectangular multi-junction solar cells made of gallium arsenide (GaAs) and other semiconductor materials.

- Specialty Power Systems

  Photovoltaics may also be incorporated as energy conversion devices for objects at elevated temperatures and with preferable radiative emissivities such as heterogeneous combustors.

Advantages

The 122 PW of sunlight reaching the Earth's surface is plentiful—almost 10,000 times more than the 13 TW equivalent of average power consumed in 2005 by humans. This abundance leads to the suggestion that it will not be long before solar energy will become the world's primary energy source. Additionally, solar electric generation has the highest power density (global mean of 170 W/m^2) among renewable energies.

Solar power is pollution-free during use. Production end-wastes and emissions are manageable using existing pollution controls. End-of-use recycling technologies are under development and policies are being produced that encourage recycling from producers.

PV installations can operate for 100 years or even more with little maintenance or intervention after their initial set-up, so after the initial capital cost of building any solar power plant, operating costs are extremely low compared to existing power technologies.

Grid-connected solar electricity can be used locally thus reducing transmission/distribution losses (transmission losses in the US were approximately 7.2% in 1995).

Compared to fossil and nuclear energy sources, very little research money has been invested in the development of solar cells, so there is considerable room for improvement. Nevertheless, experimental high efficiency solar cells already have efficiencies of over 40% in case of concentrating photovoltaic cells and efficiencies are rapidly rising while mass-production costs are rapidly falling.

In some states of the United States, much of the investment in a home-mounted system may be lost if the home-owner moves and the buyer puts less value on the system than the seller. The city of Berkeley developed an innovative financing method to remove this limitation, by adding a tax assessment that is transferred with the home to pay for the solar panels. Now known as PACE, Property Assessed Clean Energy, 30 U.S. states have duplicated this solution.

There is evidence, at least in California, that the presence of a home-mounted solar system can actually increase the value of a home. According to a paper published in April 2011 by the Ernest Orlando Lawrence Berkeley National Laboratory titled An Analysis of the Effects of Residential Photovoltaic Energy Systems on Home Sales Prices in California:

The research finds strong evidence that homes with PV systems in California have sold for a premium over comparable homes without PV systems. More specifically, estimates for average PV premiums range from approximately $3.9 to $6.4 per installed watt (DC) among a large number of different model specifications, with most models coalescing near $5.5/watt. That value corresponds to a premium of approximately $17,000 for a relatively new 3,100 watt PV system (the average size of PV systems in the study).

References

- Carlos Hernando Díaz, Sung-Mo Kang, Charvaka Duvvury, Modeling of electrical overstress in integrated circuits Springer, 1995 ISBN 0-7923-9505-0 page 5
- Downie, Neil A., Exploding Disk Cannons, Slimemobiles and 32 Other Projects for Saturday Science (Johns Hopkins University Press (2006), ISBN 978-0-8018-8506-8.
- Ahmad Y Hassan, Donald Routledge Hill (1986). Islamic Technology: An illustrated history, p. 54. Cambridge University Press. ISBN 0-521-42239-6.
- Morthorst, Poul Erik; Redlinger, Robert Y.; Andersen, Per (2002). Wind energy in the 21st century: economics, policy, technology and the changing electricity industry. Houndmills, Basingstoke, Hampshire: Palgrave/UNEP. ISBN 0-333-79248-3.
- Palz, Wolfgang (2013). Solar Power for the World: What You Wanted to Know about Photovoltaics. CRC Press. pp. 131–. ISBN 978-981-4411-87-5.
- Luque, Antonio & Hegedus, Steven (2003). Handbook of Photovoltaic Science and Engineering. John Wiley and Sons. ISBN 0-471-49196-9.
- Morten Lund (30 May 2016). "Dansk firma sætter prisbelønnet selvhejsende kran i serieproduktion". Ingeniøren. Retrieved 30 May 2016.
- Tom Gray (11 March 2013). "Fact check: About those 'abandoned' turbines ...". American Wind Energy Association. Retrieved 30 May 2016.
- Frank, Dimroth. "New world record for solar cell efficiency at 46% French-German cooperation confirms competitive advantage of European photovoltaic industry". Fraunhofer-Gesellschaft. Retrieved 14 March 2016.
- "Global Market Outlook for Photovoltaics 2014–2018" (PDF). EPIA – European Photovoltaic Industry Association. p. 45. Archived from the original on 12 June 2014. Retrieved 19 May 2015.
- Liebreich, Michael (29 January 2014). "A YEAR OF CRACKING ICE: 10 PREDICTIONS FOR 2014". Bloomberg New Energy Finance. Retrieved 24 April 2014.
- "2014 Outlook: Let the Second Gold Rush Begin" (PDF). Deutsche Bank Markets Research. 6 January 2014. Archived from the original on 22 November 2014. Retrieved 22 November 2014.

4

Electric Power System: An Overview

An electric power system, sometimes referred to as the grid consists of a network of electronic units that are used to supply, transfer and use electricity. In this chapter the reader is introduced to the key concepts of electric power system like electric power distribution, electric power transmission, power-system protection, transformer and fault current limiter. By focusing on these topics the content aims at increasing the reader's awareness about the complex topics of electric power system.

Electric Power System

An electric power system is a network of electrical components used to supply, transfer and use electric power. An example of an electric power system is the network that supplies a region's homes and industry with power—for sizeable regions, this power system is known as the grid and can be broadly divided into the generators that supply the power, the transmission system that carries the power from the generating centres to the load centres and the distribution system that feeds the power to nearby homes and industries. Smaller power systems are also found in industry, hospitals, commercial buildings and homes. The majority of these systems rely upon three-phase AC power—the standard for large-scale power transmission and distribution across the modern world. Specialised power systems that do not always rely upon three-phase AC power are found in aircraft, electric rail systems, ocean liners and automobiles.

History

That same year in London Lucien Gaulard and John Dixon Gibbs demonstrated the first transformer suitable for use in a real power system. The practical value of Gaulard and Gibbs' transformer was demonstrated in 1884 at Turin where the transformer was used to light up forty kilometres (25 miles) of railway from a single alternating current generator. Despite the success of the system, the pair made some fundamental mistakes. Perhaps the most serious was connecting the primaries of the transformers in series so that active lamps would affect the brightness of other lamps further down the line.

In 1885 George Westinghouse, an American entrepreneur, obtained the patent rights to the Gaulard Gibbs transformer and imported a number of them along with a Siemens generator and set his engineers to experimenting with them in the hopes of improving them for use in a commercial power system. One of Westinghouse's engineers, William

Stanley, recognised the problem with connecting transformers in series as opposed to parallel and also realised that making the iron core of a transformer a fully enclosed loop would improve the voltage regulation of the secondary winding. Using this knowledge he built the first practical transformer based alternating current power system at Great Barrington, Massachusetts in 1886. Westinghouse would begin installing multi-voltage AC transformer systems in competition with the Edison company later that year. In 1888 Westinghouse would also licensed Nikola Tesla's US patents for a polyphase AC induction motor and transformer designs and hired Tesla for one year to be a consultant at the Westinghouse Electric & Manufacturing Company's Pittsburgh labs.

By 1888 the electric power industry was flourishing, and power companies had built thousands of power systems (both direct and alternating current) in the United States and Europe. These networks were effectively dedicated to providing electric lighting. During this time the rivalry between Thomas Edison and George Westinghouse's companies had grown into propaganda campaign over which form of transmission (direct or alternating current) was superior, a searies of events known as the "War of Currents". In 1891, Westinghouse installed the first major power system that was designed to drive a 100 horsepower (75 kW) synchronous electric motor, not just provide electric lighting, at Telluride, Colorado. On the other side of the Atlantic, Mikhail Dolivo-Dobrovolsky built a 20 kV 176 km three-phase transmission line from Lauffen am Neckar to Frankfurt am Main for the Electrical Engineering Exhibition in Frankfurt. In the US the AC/DC competition came to the end when Edison General Electric was taken over by their chief AC rival, the Thomson-Houston Electric Company, forming General Electric. In 1895, after a protracted decision-making process, alternating current was chosen as the transmission standard with Westinghouse building the Adams No. 1 generating station at Niagara Falls and General Electric building the three-phase alternating current power system to supply Buffalo at 11 kV.

Developments in power systems continued beyond the nineteenth century. In 1936 the first experimental HVDC (high voltage direct current) line using mercury arc valves was built between Schenectady and Mechanicville, New York. HVDC had previously been achieved by series-connected direct current generators and motors (the Thury system) although this suffered from serious reliability issues. In 1957 Siemens demonstrated the first solid-state rectifier, but it was not until the early 1970s that solid-state devices became the standard in HVDC. In recent times, many important developments have come from extending innovations in the ICT field to the power engineering field. For example, the development of computers meant load flow studies could be run more efficiently allowing for much better planning of power systems. Advances in information technology and telecommunication also allowed for remote control of a power system's switchgear and generators.

Basics of Electric Power

Electric power is the product of two quantities: current and voltage. These two quantities can vary with respect to time (AC power) or can be kept at constant levels (DC power).

Most refrigerators, air conditioners, pumps and industrial machinery use AC power whereas most computers and digital equipment use DC power (the digital devices you plug into the mains typically have an internal or external power adapter to convert from AC to DC power). AC power has the advantage of being easy to transform between voltages and is able to be generated and utilised by brushless machinery. DC power remains the only practical choice in digital systems and can be more economical to transmit over long distances at very high voltages.

The ability to easily transform the voltage of AC power is important for two reasons: Firstly, power can be transmitted over long distances with less loss at higher voltages. So in power systems where generation is distant from the load, it is desirable to step-up (increase) the voltage of power at the generation point and then step-down (decrease) the voltage near the load. Secondly, it is often more economical to install turbines that produce higher voltages than would be used by most appliances, so the ability to easily transform voltages means this mismatch between voltages can be easily managed.

Solid state devices, which are products of the semiconductor revolution, make it possible to transform DC power to different voltages, build brushless DC machines and convert between AC and DC power. Nevertheless, devices utilising solid state technology are often more expensive than their traditional counterparts, so AC power remains in widespread use.

Balancing the Grid

One of the main difficulties in power systems is that the amount of active power consumed plus losses should always equal the active power produced. If more power would be produced than consumed the frequency would rise and vice versa. Even small deviations from the nominal frequency value would damage synchronous machines and other appliances. Making sure the frequency is constant is usually the task of a transmission system operator. In some countries (for example in the European Union) this is achieved through a balancing market using ancillary services.

Components of Power Systems

Supplies

All power systems have one or more sources of power. For some power systems, the source of power is external to the system but for others it is part of the system itself—it is these internal power sources that are discussed in the remainder of this section. Direct current power can be supplied by batteries, fuel cells or photovoltaic cells. Alternating current power is typically supplied by a rotor that spins in a magnetic field in a device known as a turbo generator. There have been a wide range of techniques used to spin a turbine's rotor, from steam heated using fossil fuel (including coal, gas and oil) or nuclear energy, falling water (hydroelectric power) and wind (wind power).

The speed at which the rotor spins in combination with the number of generator poles determines the frequency of the alternating current produced by the generator. All generators on a single synchronous system, for example the national grid, rotate at sub-multiples of the same speed and so generate electric current at the same frequency. If the load on the system increases, the generators will require more torque to spin at that speed and, in a typical power station, more steam must be supplied to the turbines driving them. Thus the steam used and the fuel expended are directly dependent on the quantity of electrical energy supplied. An exception exists for generators incorporating power electronics such as gearless wind turbines or linked to a grid through an asynchronous tie such as a HVDC link – these can operate at frequencies independent of the power system frequency.

The majority of the world's power still comes from coal-fired power stations like this.

Depending on how the poles are fed, alternating current generators can produce a variable number of phases of power. A higher number of phases leads to more efficient power system operation but also increases the infrastructure requirements of the system.

Electricity grid systems connect multiple generators and loads operating at the same frequency and number of phases, the commonest being three-phase at 50 or 60 Hz. However, there are other considerations. These range from the obvious: How much power should the generator be able to supply? What is an acceptable length of time for starting the generator (some generators can take hours to start)? Is the availability of the power source acceptable (some renewables are only available when the sun is shining or the wind is blowing)? To the more technical: How should the generator start (some turbines act like a motor to bring themselves up to speed in which case they need an appropriate starting circuit)? What is the mechanical speed of operation for the turbine and consequently what are the number of poles required? What type of generator is suitable (synchronous or asynchronous) and what type of rotor (squirrel-cage rotor, wound rotor, salient pole rotor or cylindrical rotor)?

Loads

Power systems deliver energy to loads that perform a function. These loads range from household appliances to industrial machinery. Most loads expect a certain voltage and, for alternating current devices, a certain frequency and number of phases. The appliances found in your home, for example, will typically be single-phase operating at 50 or 60 Hz with a voltage between 110 and 260 volts (depending on national standards). An exception exists for centralized air conditioning systems as these are now typically three-phase because this allows them to operate more efficiently. All devices in your house will also have a wattage, this specifies the amount of power the device consumes. At any one time, the net amount of power consumed by the loads on a power system must equal the net amount of power produced by the supplies less the power lost in transmission.

A toaster is great example of a single-phase load that might appear in a residence. Toasters typically draw 2 to 10 amps at 110 to 260 volts consuming around 600 to 1200 watts of power

Making sure that the voltage, frequency and amount of power supplied to the loads is in line with expectations is one of the great challenges of power system engineering. However it is not the only challenge, in addition to the power used by a load to do useful work (termed real power) many alternating current devices also use an additional amount of power because they cause the alternating voltage and alternating current to become slightly out-of-sync (termed reactive power). The reactive power like the real power must balance (that is the reactive power produced on a system must equal the reactive power consumed) and can be supplied from the generators, however it is often more economical to supply such power from capacitors.

A final consideration with loads is to do with power quality. In addition to sustained overvoltages and undervoltages (voltage regulation issues) as well as sustained deviations from the system frequency (frequency regulation issues), power system loads can be adversely affected by a range of temporal issues. These include voltage sags, dips and swells, transient overvoltages, flicker, high frequency noise, phase imbalance and poor power factor. Power quality issues occur when the power supply to a load deviates from the ideal: For an AC supply, the ideal is the current and voltage in-sync fluctuating as a perfect sine wave at a prescribed frequency with the voltage at a prescribed amplitude. For DC supply, the ideal is the voltage not varying from a prescribed level. Power quality issues can be especially important when it comes to specialist industrial machinery or hospital equipment.

Conductors

Conductors carry power from the generators to the load. In a grid, conductors may be classified as belonging to the transmission system, which carries large amounts of power at high voltages (typically more than 69 kV) from the generating centres to the load centres, or the distribution system, which feeds smaller amounts of power at lower voltages (typically less than 69 kV) from the load centres to nearby homes and industry.

Choice of conductors is based upon considerations such as cost, transmission losses and other desirable characteristics of the metal like tensile strength. Copper, with lower resistivity than Aluminum, was the conductor of choice for most power systems. However, Aluminum has lower cost for the same current carrying capacity and is the primary metal used for transmission line conductors. Overhead line conductors may be reinforced with steel or aluminum alloys.

Conductors in exterior power systems may be placed overhead or underground. Overhead conductors are usually air insulated and supported on porcelain, glass or polymer insulators. Cables used for underground transmission or building wiring are insulated with cross-linked polyethylene or other flexible insulation. Large conductors are stranded for ease of handling; small conductors used for building wiring are often solid, especially in light commercial or residential construction.

Conductors are typically rated for the maximum current that they can carry at a given temperature rise over ambient conditions. As current flow increases through a conductor it heats up. For insulated conductors, the rating is determined by the insulation. For overhead conductors, the rating is determined by the point at which the sag of the conductors would become unacceptable.

Capacitors and Reactors

The majority of the load in a typical AC power system is inductive; the current lags behind the voltage. Since the voltage and current are out-of-phase, this leads to the emergence of an "imaginary" form of power known as reactive power. Reactive power does no measurable work but is transmitted back and forth between the reactive power source and load every cycle. This reactive power can be provided by the generators themselves, through the adjustment of generator excitation, but it is often cheaper to provide it through capacitors, hence capacitors are often placed near inductive loads to reduce current demand on the power system (i.e., increase the power factor), which may never exceed 1.0, and which represents a purely resistive load. Power factor correction may be applied at a central substation, through the use of so-called "synchronous condensers" (synchronous machines which act as condensers which are variable in VAR value, through the adjustment of machine excitation) or adjacent to large loads, through the use of so-called "static condensers" (condensers which are fixed in VAR value).

Reactors consume reactive power and are used to regulate voltage on long transmission lines. In light load conditions, where the loading on transmission lines is well below the surge impedance loading, the efficiency of the power system may actually be improved by switching in reactors. Reactors installed in series in a power system also limit rushes of current flow, small reactors are therefore almost always installed in series with capacitors to limit the current rush associated with switching in a capacitor. Series reactors can also be used to limit fault currents.

Capacitors and reactors are switched by circuit breakers, which results in moderately large steps in reactive power. A solution comes in the form of static VAR compensators and static synchronous compensators. Briefly, static VAR compensators work by switching in capacitors using thyristors as opposed to circuit breakers allowing capacitors to be switched-in and switched-out within a single cycle. This provides a far more refined response than circuit breaker switched capacitors. Static synchronous compensators take a step further by achieving reactive power adjustments using only power electronics.

Power Electronics

Power electronics are semi-conductor based devices that are able to switch quantities of power ranging from a few hundred watts to several hundred megawatts. Despite their relatively simple function, their speed of operation (typically in the order of nanoseconds) means they are capable of a wide range of tasks that would be difficult or impossible with conventional technology. The classic function of power electronics is rectification, or the conversion of AC-to-DC power, power electronics are therefore found in almost every digital device that is supplied from an AC source either as an adapter that plugs into the wall or as component internal to the device. High-powered power electronics can also be used to convert AC power to DC power for long distance transmission in a system known as HVDC. HVDC is used because it proves to be more economical than similar high voltage AC systems for very long distances (hundreds to thousands of kilometres). HVDC is also desirable for interconnects because it allows frequency independence thus improving system stability. Power electronics are also essential for any power source that is required to produce an AC output but that by its nature produces a DC output. They are therefore used by many photovoltaic installations both industrial and residential.

Power electronics also feature in a wide range of more exotic uses. They are at the heart of all modern electric and hybrid vehicles—where they are used for both motor control and as part of the brushless DC motor. Power electronics are also found in practically all modern petrol-powered vehicles, this is because the power provided by the car's batteries alone is insufficient to provide ignition, air-conditioning, internal lighting, radio and dashboard displays for the life of the car. So the batteries must be recharged while driving using DC power from the engine—a feat that is typically accomplished using power electronics. Whereas conventional technology would be unsuitable for a modern electric car, commutators can and have been used in petrol-powered cars, the switch to alternators in combination with power electronics has occurred because of the improved durability of brushless machinery.

Some electric railway systems also use DC power and thus make use of power electronics to feed grid power to the locomotives and often for speed control of the locomotive's motor. In the middle twentieth century, rectifier locomotives were popular, these used power electronics to convert AC power from the railway network for use by a DC motor. Today most electric locomotives are supplied with AC power and run using AC motors, but still use power electronics to provide suitable motor control. The use of power electronics to assist with motor control and with starter circuits cannot be underestimated and, in addition to rectification, is responsible for power electronics appearing in a wide range of industrial machinery. Power electronics even appear in modern residential air conditioners.

Power electronics are also at the heart of the variable speed wind turbine. Conventional wind turbines require significant engineering to ensure they operate at some ratio of the system frequency, however by using power electronics this requirement can be eliminated leading to quieter, more flexible and (at the moment) more costly wind turbines. A final example of one of the more exotic uses of power electronics comes from the previous section where the fast-switching times of power electronics were used to provide more refined reactive compensation to the power system.

Protective Devices

Power systems contain protective devices to prevent injury or damage during failures. The quintessential protective device is the fuse. When the current through a fuse exceeds a certain threshold, the fuse element melts, producing an arc across the resulting gap that is then extinguished, interrupting the circuit. Given that fuses can be built as the weak point of a system, fuses are ideal for protecting circuitry from damage. Fuses however have two problems: First, after they have functioned, fuses must be replaced as they cannot be reset. This can prove inconvenient if the fuse is at a remote site or a spare fuse is not on hand. And second, fuses are typically inadequate as the sole safety device in most power systems as they allow current flows well in excess of that that would prove lethal to a human or animal.

The first problem is resolved by the use of circuit breakers—devices that can be reset after they have broken current flow. In modern systems that use less than about 10 kW, miniature circuit breakers are typically used. These devices combine the mechanism that initiates the trip (by sensing excess current) as well as the mechanism that breaks the current flow in a single unit. Some miniature circuit breakers operate solely on the basis of electromagnetism. In these miniature circuit breakers, the current is run through a solenoid, and, in the event of excess current flow, the magnetic pull of the solenoid is sufficient to force open the circuit breaker's contacts (often indirectly through a tripping mechanism). A better design however arises by inserting a bimetallic strip before the solenoid—this means that instead of always producing a magnetic force, the solenoid only produces a magnetic force when the current is strong enough to deform the bimetallic strip and complete the solenoid's circuit.

In higher powered applications, the protective relays that detect a fault and initiate a trip are separate from the circuit breaker. Early relays worked based upon electromagnetic principles similar to those mentioned in the previous paragraph, modern relays are application-specific computers that determine whether to trip based upon readings from the power system. Different relays will initiate trips depending upon different protection schemes. For example, an overcurrent relay might initiate a trip if the current on any phase exceeds a certain threshold whereas a set of differential relays might initiate a trip if the sum of currents between them indicates there may be current leaking to earth. The circuit breakers in higher powered applications are different too. Air is typically no longer sufficient to quench the arc that forms when the contacts are forced open so a variety of techniques are used. One of the most popular techniques is to keep the chamber enclosing the contacts flooded with sulfur hexafluoride (SF_6)—a non-toxic gas that has sound arc-quenching properties. Other techniques are discussed in the reference.

The second problem, the inadequacy of fuses to act as the sole safety device in most power systems, is probably best resolved by the use of residual current devices (RCDs). In any properly functioning electrical appliance the current flowing into the appliance on the active line should equal the current flowing out of the appliance on the neutral line. A residual current device works by monitoring the active and neutral lines and tripping the active line if it notices a difference. Residual current devices require a separate neutral line for each phase and to be able to trip within a time frame before harm occurs. This is typically not a problem in most residential applications where standard wiring provides an active and neutral line for each appliance (that's why your power plugs always have at least two tongs) and the voltages are relatively low however these issues do limit the effectiveness of RCDs in other applications such as industry. Even with the installation of an RCD, exposure to electricity can still prove lethal.

SCADA Systems

In large electric power systems, Supervisory Control And Data Acquisition (SCADA) is used for tasks such as switching on generators, controlling generator output and switching in or out system elements for maintenance. The first supervisory control systems implemented consisted of a panel of lamps and switches at a central console near the controlled plant. The lamps provided feedback on the state of plant (the data acquisition function) and the switches allowed adjustments to the plant to be made (the supervisory control function). Today, SCADA systems are much more sophisticated and, due to advances in communication systems, the consoles controlling the plant no longer need to be near the plant itself. Instead it is now common for plants to be controlled with equipment similar (if not identical) to a desktop computer. The ability to control such plants through computers has increased the need for security—there have already been reports of cyber-attacks on such systems causing significant disruptions to power systems.

Power Systems in Practice

Despite their common components, power systems vary widely both with respect to their design and how they operate. This section introduces some common power system types and briefly explains their operation.

Residential Power Systems

Residential dwellings almost always take supply from the low voltage distribution lines or cables that run past the dwelling. These operate at voltages of between 110 and 260 volts (phase-to-earth) depending upon national standards. A few decades ago small dwellings would be fed a single phase using a dedicated two-core service cable (one core for the active phase and one core for the neutral return). The active line would then be run through a main isolating switch in the fuse box and then split into one or more circuits to feed lighting and appliances inside the house. By convention, the lighting and appliance circuits are kept separate so the failure of an appliance does not leave the dwelling's occupants in the dark. All circuits would be fused with an appropriate fuse based upon the wire size used for that circuit. Circuits would have both an active and neutral wire with both the lighting and power sockets being connected in parallel. Sockets would also be provided with a protective earth. This would be made available to appliances to connect to any metallic casing. If this casing were to become live, the theory is the connection to earth would cause an RCD or fuse to trip—thus preventing the future electrocution of an occupant handling the appliance. Earthing systems vary between regions, but in countries such as the United Kingdom and Australia both the protective earth and neutral line would be earthed together near the fuse box before the main isolating switch and the neutral earthed once again back at the distribution transformer.

There have been a number of minor changes over the year to practice of residential wiring. Some of the most significant ways modern residential power systems tend to vary from older ones include:

- For convenience, miniature circuit breakers are now almost always used in the fuse box instead of fuses as these can easily be reset by occupants.
- For safety reasons, RCDs are now installed on appliance circuits and, increasingly, even on lighting circuits.
- Dwellings are typically connected to all three-phases of the distribution system with the phases being arbitrarily allocated to the house's single-phase circuits.

- Whereas air conditioners of the past might have been fed from a dedicated circuit attached to a single phase, centralised air conditioners that require three-phase power are now becoming common.
- Protective earths are now run with lighting circuits to allow for metallic lamp holders to be earthed.
- Increasingly residential power systems are incorporating microgenerators, most notably, photovoltaic cells.

Commercial Power Systems

Commercial power systems such as shopping centers or high-rise buildings are larger in scale than residential systems. Electrical designs for larger commercial systems are usually studied for load flow, short-circuit fault levels, and voltage drop for steady-state loads and during starting of large motors. The objectives of the studies are to assure proper equipment and conductor sizing, and to coordinate protective devices so that minimal disruption is cause when a fault is cleared. Large commercial installations will have an orderly system of sub-panels, separate from the main distribution board to allow for better system protection and more efficient electrical installation.

Typically one of the largest appliances connected to a commercial power system is the HVAC unit, and ensuring this unit is adequately supplied is an important consideration in commercial power systems. Regulations for commercial establishments place other requirements on commercial systems that are not placed on residential systems. For example, in Australia, commercial systems must comply with AS 2293, the standard for emergency lighting, which requires emergency lighting be maintained for at least 90 minutes in the event of loss of mains supply. In the United States, the National Electrical Code requires commercial systems to be built with at least one 20A sign outlet in order to light outdoor signage. Building code regulations may place special requirements on the electrical system for emergency lighting, evacuation, emergency power, smoke control and fire protection.

Electric Power Distribution

An electric power distribution system is the final stage in the delivery of electric power; it carries electricity from the transmission system to individual consumers. Distribution substations connect to the transmission system and lower the transmission voltage to medium voltage ranging between 2 kV and 35 kV with the use of transformers. Primary distribution lines carry this medium voltage power to distribution transformers located near the customer's premises. Distribution transformers again lower the voltage to the utilization voltage of household appliances and typically feed several customers through secondary distribution lines at this voltage. Commercial and residential customers are connected to the secondary distribution lines through service drops. Customers demanding a much larger amount of power may be connected directly to the primary distribution level or the subtransmission level.

A 50 kVA pole-mounted distribution transformer

History

Electric power distribution only became necessary in the 1880s when electricity started being generated at power stations. Before that electricity was usually generated where it was used. The first power distribution systems installed in European and US cites were used to supply lighting: arc lighting running on very high voltage (usually higher than 3000 volt) alternating current (AC) or direct current (DC), and incandescent lighting running on low voltage (100 volt) direct current. Both were supplanting gas lighting systems, with arc lighting taking over large area/street lighting, and incandescent lighting replacing gas for business and residential lighting.

The late 1870s/early 1880s saw the introduction of arc lamp lighting used outdoors or in large indoor spaces such as this Brush Electric Company system installed in 1880 in New York City.

Due to the high voltages used in arc lighting, a single generating station could supply a long string of lights, up to 7-mile (11 km) long circuits, since the capacity of a wire is proportional to the square of the current traveling on it, each doubling of the voltage would allow the same size cable to transmit the same amount of power four times the distance. Direct current indoor incandescent lighting systems (for example the first Edison Pearl Street Station installed in 1882), had difficulty supplying customers more than a mile away due to the low 110 volt system being used throughout the system, from the generators to the final use. The Edison DC system needed thick copper conductor cables, and the generating plants needed to be within about 1.5 miles (2.4 km) of the farthest customer to avoid excessively large and expensive conductors.

Introduction of the AC Transformer

Trying to deliver electricity long distance at high voltage and then reducing it to a fractional voltage for indoor lighting became a recognized engineering roadblock to electric power distribution with many, not very satisfactory, solutions tested by lighting companies. The mid-1880s saw a breakthrough with the development of functional AC transformers that allowed the voltage to be "stepped up" to much higher transmission voltages and then dropped down to a lower end user voltage. With much cheaper transmission costs and the greater economies of scale of having large generating plants supply whole cities and regions, the use of AC spread rapidly.

In the US the competition between direct current and alternating current took a personal turn in the late 1880s in the form of a "War of Currents" when Thomas Edison started attacking George Westinghouse and his development of the first US AC transformer systems, pointing out all the deaths caused by high voltage AC systems over the years and claiming any AC system was inherently dangerous. Edison's propaganda campaign was short lived with his company switching over to AC in 1892.

AC became the dominant form of transmission of power with innovations in Europe and the US in electric motor designs and the development of engineered universal systems allowing the large number of legacy systems to be connected to large AC grids.

In the first half of the 20th century, the electric power industry was vertically integrated, meaning that one company did generation, transmission, distribution, metering and billing. Starting in the 1970s and 1980s nations began the process of deregulation and privatisation, leading to electricity markets. The distribution system would remain regulated, but generation, retail, and sometimes transmission systems were transformed into competitive markets.

Generation and Transmission

Electric power begins at a generating station, where the potential difference can be as high as 13,800 volts. AC is usually used. Users of large amounts of DC power such as some railway electrification systems, telephone exchanges and industrial processes such as aluminium smelting usually either operate their own or have adjacent dedicated generating equipment, or use rectifiers to derive DC from the public AC supply. However, High-voltage DC can be advantageous for isolating alternating-current systems or controlling the quantity of electricity transmitted. For example, Hydro-Québec has a direct-current line which goes from the James Bay region to Boston.

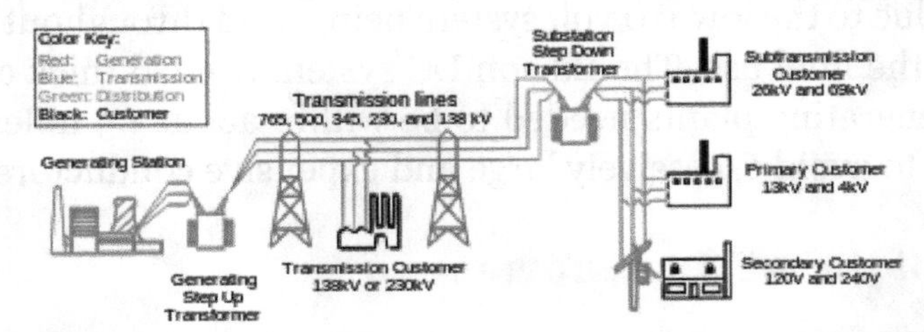

Simplified diagram of AC electricity delivery from generation stations to consumers' service drop.

From the generating station it goes to the generating station's switchyard where a step-up transformer increases the voltage to a level suitable for transmission, from 44kV to 765kV. Once in the transmission system, electricity from each generating station is combined with electricity produced elsewhere. Electricity is consumed as soon as it is produced. It is transmitted at a very high speed, close to the speed of light.

Distribution Overview

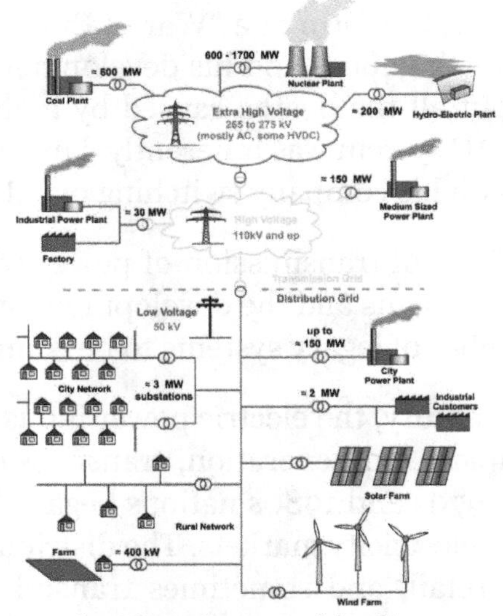

General layout of electricity networks. The voltages and loadings are typical of a European network.

The transition from transmission to distribution happens in a power substation, which has the following functions:

- Circuit breakers and switches enable the substation to be disconnected from the transmission grid or for distribution lines to be disconnected.

- Transformers step down transmission voltages, 35kV or more, down to primary distribution voltages. These are medium voltage circuits, usually 600-35,000 V.
- From the transformer, power goes to the busbar that can split the distribution power off in multiple directions. The bus distributes power to distribution lines, which fan out to customers.

Urban distribution is mainly underground, sometimes in common utility ducts. Rural distribution is mostly above ground with utility poles, and suburban distribution is a mix. Closer to the customer, a distribution transformer steps the primary distribution power down to a low-voltage secondary circuit, usually 120 or 240V, depending on the region. The power comes to the customer via a service drop and an electricity meter. The final circuit in an urban system may be less than 50 feet, but may be over 300 feet for a rural customer.

Primary Distribution

Primary distribution voltages are 22kV or 11 kV. Only large consumers are fed directly from distribution voltages; most utility customers are connected to a transformer, which reduces the distribution voltage to the low voltage used by lighting and interior wiring systems.

Voltage varies according to its role in the supply and distribution system. According to international standards, there are initially two voltage groups: low voltage (LV): up to and including 1kV AC (or 1.5kV DC) and high voltage (HV): above 1 kV AC (or 1.5 kV DC).

Network Configurations

Distribution networks are divided into two types, radial or network. A radial system is arranged like a tree where each customer has one source of supply. A network system has multiple sources of supply operating in parallel. Spot networks are used for concentrated loads. Radial systems are commonly used in rural or suburban areas.

Substation near Yellowknife, in the Northwest Territories of Canada

Radial systems usually include emergency connections where the system can be reconfigured in case of problems, such as a fault or required replacement. This can be done by opening and closing switches. It may be acceptable to close a loop for a short time.

Long feeders experience voltage drop (power factor distortion) requiring capacitors to be installed.

Reconfiguration, by exchanging the functional links between the elements of the system, represents one of the most important measures which can improve the operational performance of a distribution system. The problem of optimization through the reconfiguration of a power distribution system, in terms of its definition, is a historical single objective problem with constraints. Since 1975, when Merlin and Back introduced the idea of distribution system reconfiguration for active power loss reduction, until nowadays, a lot of researchers have proposed diverse methods and algorithms to solve the reconfiguration problem as a single objective problem. Some authors have proposed Pareto optimality based approaches (including active power losses and reliability indices as objectives). For this purpose, different artificial intelligence based methods have been used: microgenetic, branch exchange, particle swarm optimization and non-dominated sorting genetic algorithm.

Rural Services

Rural electrification systems tend to use higher distribution voltages because of the longer distances covered by distribution lines. 7.2, 12.47, 25, and 34.5 kV distribution is common in the United States; 11 kV and 33 kV are common in the UK, Australia and New Zealand; 11 kV and 22 kV are common in South Africa. Other voltages are occasionally used. Distribution in rural areas may be only single-phase if it is not economical to install three-phase power for relatively few and small customers.

Rural services normally try to minimize the number of poles and wires. Single-wire earth return (SWER) is the least expensive, with one wire. It uses higher voltages (than urban distribution), which in turn permits use of galvanized steel wire. The strong steel wire allows for less expensive wide pole spacing. In rural areas a pole-mount transformer may serve only one customer.

Higher voltage split-phase or three phase service, at a higher infrastructure and a higher cost, provide increased equipment efficiency and lower energy cost for large agricultural facilities, petroleum pumping facilities, or water plants.

In New Zealand, Australia, Saskatchewan, Canada, and South Africa, single wire earth return systems (SWER) are used to electrify remote rural areas.

Secondary Distribution

Electricity is delivered at a frequency of either 50 or 60 Hz, depending on the region. It is delivered to domestic customers as single-phase electric power In some countries as in Europe a three phase supply may be made available for larger properties. Seen in an oscilloscope, the domestic power supply in North America would look like a sine wave, oscillating between -170 volts and 170 volts, giving an effective voltage of 120 volts. Three-phase power is more efficient in terms of power delivered per cable used, and is more suited to running large electric motors. Some large European appliances may be powered by three-phase power, such as electric stoves and clothes dryers.

A ground connection is normally provided for the customer's system as well as for the equipment owned by the utility. The purpose of connecting the customer's system to ground is to limit the voltage that may develop if high voltage conductors fall down onto lower-voltage conductors which are usually mounted lower to the ground, or if a failure occurs within a distribution transformer. Earthing systems can be TT, TN-S, TN-C-S or TN-C.

Regional Variations

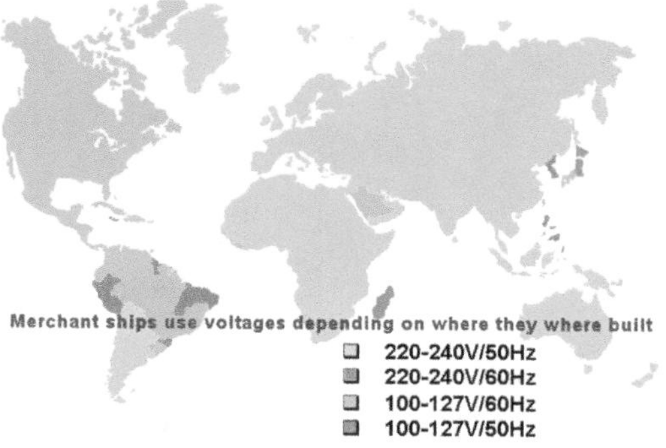

Worldwide electrical voltage and frequency.

220-240 Volt Systems

Most of the world uses 50 Hz 220 or 230 V single phase 400V 3 phase for residential and light industrial services. In this system, the primary distribution network supplies a few substations per area, and the 230 V power from each substation is directly distributed. A live (hot) wire and neutral are connected to the building for each phase of three phase service. Single-phase distribution is used where motor loads are light. In Europe, electricity is normally distributed for industry and domestic use by the three-phase, four wire system. This gives a three-phase voltage of 400 volts wye service and a single-phase voltage of 230 volts. In the UK a typical urban or suburban low-voltage substation would normally be rated between 150 kVA and 1 MVA and supply a whole neighborhood of a few hundred houses. For industrial customers, 3-phase 690 / 400 volt is also available, or may be generated locally.. Large industrial customers have their own transformer(s) with an input from 11 kV to 220 kV.

110-120 Volt Systems

Most of the Americas use 60 Hz AC, the 120/240 volt split phase system domestically and three phase for larger installations. Compared to European systems, North American ones have more step-down transformers near customers. This is because the higher domestic voltage used in Europe (230 V vs 120 V) may be carried over a greater distance with acceptable power loss. North American transforms usually power homes at 240 volts, similar to Europe's 230 volts. It is the split-phase that allows use of 120 volts in the home. The transformers actually provide 240 volts, not 120, so power can be carried over a greater distance.

Japan's utility frequencies are 50 Hz and 60 Hz

In the electricity sector in Japan, the standard frequencies for AC are 50 and 60 Hz. In Japan parts of the country use 50 Hz, while other parts use 60 Hz. This is a relic of the 1800s. Some local providers in Tokyo imported 50 Hz German equipment, while the local power providers in Osaka brought in 60 Hz generators from the United States. The grids grew until eventually the entire country was wired. Today the frequency is 50 Hz in Eastern Japan (including Tokyo, Yokohama, Tohoku, and Hokkaido) and 60 Hertz in Western Japan (including Nagoya, Osaka, Kyoto, Hiroshima, Shikoku, and Kyushu).

Most household appliances are made to work on either frequency. The problem of incompatibility came into the public eye when the 2011 Tōhoku earthquake and tsunami knocked out about a third of the east's capacity, and power in the west couldn't be fully shared with the east, since the country does not have a common frequency.

There are four high-voltage direct current (HVDC) converter stations that move power across Japan's AC frequency border. Shin Shinano is a back-to-back HVDC facility in Japan which forms one of four frequency changer stations that link Japan's western and eastern power grids. The other three are at Higashi-Shimizu, Minami-Fukumitsu and Sakuma Dam. Together they can move up to 1.2 GW of power east or west.

240 Volt Systems and 120 Volt Outlets

Most modern North American homes are wired to receive 240 volts from the transformer, and through the use of split-phase electrical power, can have both 120 volt receptacles and 240 volt receptacles. The 120 volts is typically used for lighting and most wall outlets. The 240 volt outlets are usually placed where the water heater and clothes dryer would go. Sometimes a 240 volt outlet is mounted in the garage for machinery or for charging an electric car.

Electric Power Transmission

Two-circuit, single-voltage power transmission line; "Bundled" 4-ways

Electric power transmission is the bulk movement of electrical energy from a generating site, such as a power plant, to an electrical substation. The interconnected lines which facilitate this movement are known as a transmission network. This is distinct from the local wiring between high-voltage substations and customers, which is typically referred to as electric power distribution. The combined transmission and distribution network is known as the "power grid" in North America, or just "the grid". In the United Kingdom, the network is known as the "National Grid".

A wide area synchronous grid, also known as an "interconnection" in North America, directly connects a large number of generators delivering AC power with the same relative frequency to a large number of consumers. For example, there are four major interconnections in North America (the Western Interconnection, the Eastern Interconnection, the Quebec Interconnection and the Electric Reliability Council of Texas (ERCOT) grid), and one large grid for most of continental Europe.

Power lines in the San Francisco Bay Area

The same relative frequency, but almost never the same relative phase as ac power interchange is a function of the phase difference between any two nodes in the network, and zero degrees difference means no power is interchanged; any phase difference up to 90 degrees is stable by the "equal area criteria"; any phase difference above 90 degrees is absolutely unstable; the interchange partners are responsible for maintaining frequency as close to the utility frequency as is practical, and the phase differences between any two nodes significantly less than 90 degrees; should 90 degrees be exceeded, a system separation is executed, and remains separated until the trouble has been corrected.

Power lines in Qatar

Historically, transmission and distribution lines were owned by the same company, but starting in the 1990s, many countries have liberalized the regulation of the electricity market in ways that have led to the separation of the electricity transmission business from the distribution business.

System

Most transmission lines are high-voltage three-phase alternating current (AC), although single phase AC is sometimes used in railway electrification systems. High-voltage direct-current (HVDC) technology is used for greater efficiency over very long distances (typically hundreds of miles). HVDC technology is also used in submarine power cables (typically longer than 30 miles (50 km)), and in the interchange of power between grids that are not mutually synchronized. HVDC links are used to stabilize large power distribution networks where sudden new loads, or blackouts, in one part of a network can result in synchronization problems and cascading failures.

Electricity is transmitted at high voltages (115 kV or above) to reduce the energy loss which occurs in long-distance transmission. Power is usually transmitted through overhead power lines. Underground power transmission has a significantly higher installation cost and greater operational limitations, but reduced maintenance costs. Underground transmission is sometimes used in urban areas or environmentally sensitive locations.

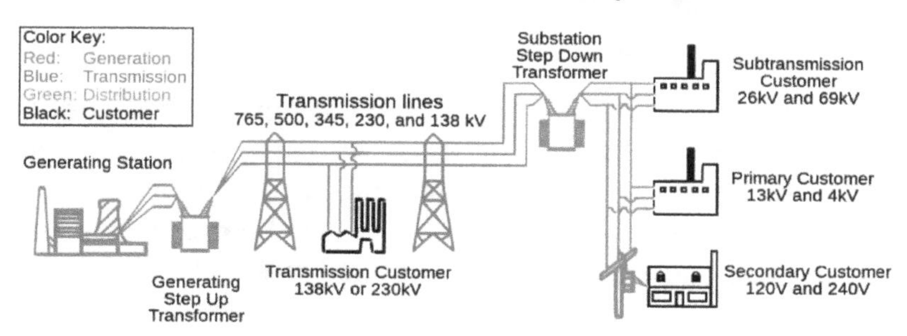

Diagram of an electric power system; transmission system is in blue

A lack of electrical energy storage facilities in transmission systems leads to a key limitation. Electrical energy must be generated at the same rate at which it is consumed. A sophisticated control system is required to ensure that the power generation very closely matches the demand. If the demand for power exceeds supply, the imbalance can cause generation plant(s) and transmission equipment to automatically disconnect and/or shut down to prevent damage. In the worst case, this may lead to a cascading series of shut downs and a major regional blackout. Examples include the US Northeast blackouts of 1965, 1977, 2003, and major blackouts in other US regions in 1996 and 2011. Electric transmission networks are interconnected into regional, national, and even continent wide networks to reduce the risk of such a failure by providing multiple redundant, alternative routes for power to flow should such shut downs occur. Transmission companies determine the maximum reliable capacity of each line (ordinarily less than its physical or thermal limit) to ensure that spare capacity is available in the event of a failure in another part of the network.

Overhead Transmission

3-phase high-voltage lines in Washington State, “Bundled” 3-ways

Four-circuit, two-voltage power transmission line; "Bundled" 2-ways

High-voltage overhead conductors are not covered by insulation. The conductor material is nearly always an aluminum alloy, made into several strands and possibly reinforced with steel strands. Copper was sometimes used for overhead transmission, but aluminum is lighter, yields only marginally reduced performance and costs much less. Overhead conductors are a commodity supplied by several companies worldwide. Improved conductor material and shapes are regularly used to allow increased capacity and modernize transmission circuits. Conductor sizes range from 12 mm² (#6 American wire gauge) to 750 mm² (1,590,000 circular mils area), with varying resistance and current-carrying capacity. Thicker wires would lead to a relatively small increase in capacity due to the skin effect, that causes most of the current to flow close to the surface of the wire. Because of this current limitation, multiple parallel cables (called bundle conductors) are used when higher capacity is needed. Bundle conductors are also used at high voltages to reduce energy loss caused by corona discharge.

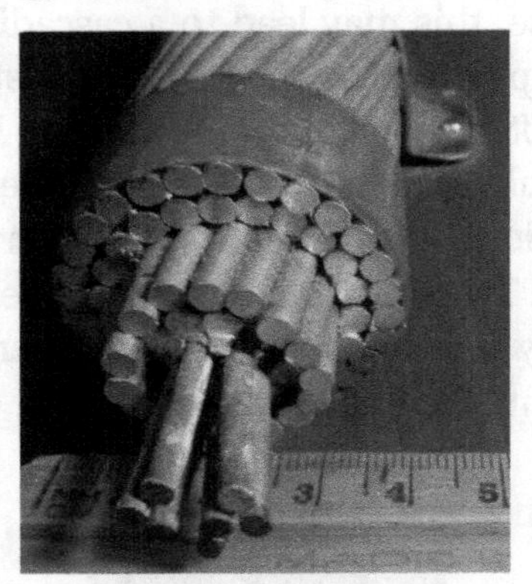

A typical ACSR. The conductor consists of seven strands of steel surrounded by four layers of aluminium.

Today, transmission-level voltages are usually considered to be 110 kV and above. Lower voltages, such as 66 kV and 33 kV, are usually considered subtransmission voltages, but are occasionally used on long lines with light loads. Voltages less than 33 kV are usually used for distribution. Voltages above 765 kV are considered extra high voltage and require different designs compared to equipment used at lower voltages.

Since overhead transmission wires depend on air for insulation, the design of these lines requires minimum clearances to be observed to maintain safety. Adverse weather conditions, such as high wind and low temperatures, can lead to power outages. Wind speeds as low as 23 knots (43 km/h) can permit conductors to encroach operating clearances, resulting in a flashover and loss of supply. Oscillatory motion of the physical line can be termed gallop or flutter depending on the frequency and amplitude of oscillation.

Underground Transmission

Electric power can also be transmitted by underground power cables instead of overhead power lines. Underground cables take up less right-of-way than overhead lines, have lower visibility, and are less affected by bad weather. However, costs of insulated cable and excavation are much higher than overhead construction. Faults in buried transmission lines take longer to locate and repair. Underground lines are strictly limited by their thermal capacity, which permits less overload or re-rating than overhead lines. Long underground AC cables have significant capacitance, which may reduce their ability to provide useful power to loads beyond 50 mi (80 km). Long underground DC cables have no such issue and can run for thousands of miles.

History

In the early days of commercial electric power, transmission of electric power at the same voltage as used by lighting and mechanical loads restricted the distance between generating plant and consumers. In 1882, generation was with direct current (DC), which could not easily be increased in voltage for long-distance transmission. Different classes of loads (for example, lighting, fixed motors, and traction/railway systems) required different voltages, and so used different generators and circuits.

New York City streets in 1890. Besides telegraph lines, multiple electric lines were required for each class of device requiring different voltages

Due to this specialization of lines and because transmission was inefficient for low-voltage high-current circuits, generators needed to be near their loads. It seemed, at the time, that the industry would develop into what is now known as a distributed generation system with large numbers of small generators located near their loads.

The transmission of electric power with alternating current (AC) became possible after Lucien Gaulard and John Dixon Gibbs built what they called the secondary generator, an early transformer provided with 1:1 turn ratio and open magnetic circuit, in 1881.

The first long distance (34 km (21 mi)) AC line was built for the 1884 International Exhibition of Turin, Italy. It was powered by a 2000-V, 130Hz Siemens & Halske alternator and featured several Gaulard secondary generators with their primary windings connected in series, which fed incandescent lamps. The system proved the feasibility of AC electric power transmission on long distances.

A very first operative AC line was put into service in 1885 in via dei Cerchi, Rome, Italy, for public lighting. It was powered by two Siemens & Halske alternators rated 30 hp (22 kW), 2000-V at 120 Hz and used 200 series-connected Gaulard 2000-V/20-V step-down transformers provided with a closed magnetic circuit, one for each lamp. Few months later it was followed by the first British AC system, which was put into service at the Grosvenor Gallery, London. It also featured Siemens alternators and 2400-V/100-V step-down transformers, one per user, with shunt-connected primaries.

Working for Westinghouse, William Stanley Jr. spent his time recovering from illness in Great Barrington installing what is considered the worlds first practical AC transformer system.

Working from what he considered an impractical Gaulard-Gibbs design, electrical engineer William Stanley, Jr. developed what is considered the first practical series AC transformer in 1885. Working with the support of George Westinghouse, in 1886 he installed demonstration transformer based alternating current lighting system in Great Barrington, Massachusetts. Powered by a steam engine driven 500-V Siemens generator, voltage was stepped down to 100 volts using the new Stanley transformer to power incandescent lamps at 23 businesses along main street with very little power loss over 4000 feet. This practical demonstration of a transformer/alternating current lighting system would lead Westinghouse to begin installing AC based systems later that year.

In 1888 alternating current systems gained further viability with introduction of a functional AC motor, something these systems had lacked up till then. The design, an induction motor running on polyphase current, was independently invented by Galileo Ferraris and Nikola Tesla (with Tesla's design being licensed by Westinghouse in the US). This design was further developed into the modern practical three-phase form by Mikhail Dolivo-Dobrovolsky and Charles Eugene Lancelot Brown.

The late 1880s/early 1890s would see a financial merger of many smaller electric companies into a few larger corporations such as Ganz and AEG in Europe and General Electric and Westinghouse Electric in the US. These companies continued to develop AC systems but the technical difference between direct and alternating current systems would follow a much longer technical merger. Due to innovation in the US and Europe, alternating current's economy of scale with very large generating plants linked to loads via long distance transmission was slowly being combined with the ability to link it up with all of the existing systems that needed to be supplied. These included single phase AC systems, poly-phase AC systems, low voltage incandescent lighting, high voltage arc lighting, and existing DC motors in factories and street cars. In what was becoming a universal system, these technological differences were temporarily being bridged via the development of rotary converters and motor–generators that would allow the large number of legacy systems to be connected to the AC grid. These stopgaps would slowly be replaced as older systems were retired or upgraded.

Westinghouse alternating current polyphase generators on display at the 1893 World's Fair in Chicago, part of their "Tesla Poly-phase System". Such polyphase innovations revolutionized transmission

The first transmission of three-phase alternating current using high voltage took place in 1891 during the international electricity exhibition in Frankfurt. A 15,000-V transmission line, approximately 175 km long, connected Lauffen on the Neckar and Frankfurt.

Voltages used for electric power transmission increased throughout the 20th century. By 1914, fifty-five transmission systems each operating at more than 70,000-V were in service. The highest voltage then used was 150,000V. By allowing multiple generating plants to be interconnected over a wide area, electricity production cost was reduced. The most efficient available plants could be used to supply the varying loads during the day. Reliability was improved and capital investment cost was reduced, since stand-by generating capacity could be shared over many more customers and a wider geographic area. Remote and low-cost sources of energy, such as hydroelectric power or mine-mouth coal, could be exploited to lower energy production cost.

The rapid industrialization in the 20th century made electrical transmission lines and grids a critical infrastructure item in most industrialized nations. The interconnection of local generation plants and small distribution networks was greatly spurred by the requirements of World War I, with large electrical generating plants built by governments to provide power to munitions factories. Later these generating plants were connected to supply civil loads through long-distance transmission.

Bulk Power Transmission

Engineers design transmission networks to transport the energy as efficiently as feasible, while at the same time taking into account economic factors, network safety and redundancy. These networks use components such as power lines, cables, circuit breakers, switches and transformers. The transmission network is usually administered on a regional basis by an entity such as a regional transmission organization or transmission system operator.

A transmission substation decreases the voltage of incoming electricity, allowing it to connect from long distance high voltage transmission, to local lower voltage distribution. It also reroutes power to other transmission lines that serve local markets. This is the PacifiCorp Hale Substation, Orem, Utah, USA

Transmission efficiency is greatly improved by devices that increase the voltage (and thereby proportionately reduce the current), in the line conductors, thus allowing power to be transmitted with acceptable losses. The reduced current flowing through the line reduces the heating losses in the conductors. According to Joule's Law, energy losses are directly proportional to the square of the current. Thus, reducing the current by a factor of two will lower the energy lost to conductor resistance by a factor of four for any given size of conductor.

The optimum size of a conductor for a given voltage and current can be estimated by Kelvin's law for conductor size, which states that the size is at its optimum when the annual cost of energy wasted in the resistance is equal to the annual capital charges of providing the conductor. At times of lower interest rates, Kelvin's law indicates that thicker wires are optimal; while, when metals are expensive, thinner conductors are indicated: however, power lines are designed for long-term use, so Kelvin's law has to be used in conjunction with long-term estimates of the price of copper and aluminum as well as interest rates for capital.

The increase in voltage is achieved in AC circuits by using a step-up transformer. HVDC systems require relatively costly conversion equipment which may be economically justified for particular projects such as submarine cables and longer distance high capacity point to point transmission. HVDC is necessary for the import and export of energy between grid systems that are not synchronized with each other.

A transmission grid is a network of power stations, transmission lines, and substations. Energy is usually transmitted within a grid with three-phase AC. Single-phase AC is used only for distribution to end users since it is not usable for large polyphase induction motors. In the 19th century, two-phase transmission was used but required either four wires or three wires with unequal currents. Higher order phase systems require more than three wires, but deliver little or no benefit.

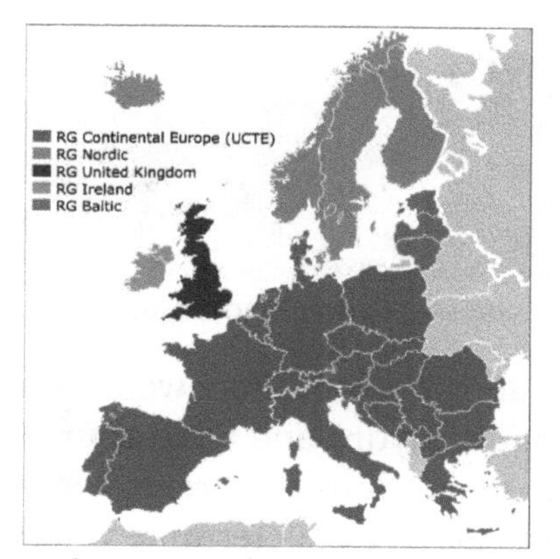

The synchronous grids of the European Union

The price of electric power station capacity is high, and electric demand is variable, so it is often cheaper to import some portion of the needed power than to generate it locally. Because loads are often regionally correlated (hot weather in the Southwest portion of the US might cause many people to use air conditioners), electric power often comes from distant sources. Because of the economic benefits of load sharing between regions, wide area transmission grids now span countries and even continents. The web of interconnections between power producers and consumers should enable power to flow, even if some links are inoperative.

The unvarying (or slowly varying over many hours) portion of the electric demand is known as the base load and is generally served by large facilities (which are more efficient due to economies of scale) with fixed costs for fuel and operation. Such facilities are nuclear, coal-fired or hydroelectric, while other energy sources such as concentrated solar thermal and geothermal power have the potential to provide base load power. Renewable energy sources, such as solar photovoltaics, wind, wave, and tidal, are, due to their intermittency, not considered as supplying "base load" but will still add power to the grid. The remaining or 'peak' power demand, is supplied by peaking power plants, which are typically smaller, faster-responding, and higher cost sources, such as combined cycle or combustion turbine plants fueled by natural gas.

Long-distance transmission of electricity (thousands of kilometers) is cheap and efficient, with costs of US$0.005–0.02/kWh (compared to annual averaged large producer costs of US$0.01–0.025/kWh, retail rates upwards of US$0.10/kWh, and multiples of retail for instantaneous suppliers at unpredicted highest demand moments). Thus distant suppliers can be cheaper than local sources (e.g., New York often buys over 1000 MW of electricity from Canada). Multiple local sources (even if more expensive and infrequently used) can make the transmission grid more fault tolerant to weather and other disasters that can disconnect distant suppliers.

A high-power electrical transmission tower, 230 kV, double-circuit, also double-bundled

Long-distance transmission allows remote renewable energy resources to be used to displace fossil fuel consumption. Hydro and wind sources cannot be moved closer to populous cities, and solar costs are lowest in remote areas where local power needs are minimal. Connection costs alone can determine whether any particular renewable alternative is economically sensible. Costs can be prohibitive for transmission lines, but various proposals for massive infrastructure investment in high capacity, very long distance super grid transmission networks could be recovered with modest usage fees.

Grid Input

At the power stations, the power is produced at a relatively low voltage between about 2.3 kV and 30 kV, depending on the size of the unit. The generator terminal voltage is then stepped up by the power station transformer to a higher voltage (115 kV to 765 kV AC, varying by the transmission system and by country) for transmission over long distances.

Losses

Transmitting electricity at high voltage reduces the fraction of energy lost to resistance, which varies depending on the specific conductors, the current flowing, and the length of the transmission line. For example, a 100-mile (160 km) 765 kV line carrying 1000

MW of power can have losses of 1.1% to 0.5%. A 345 kV line carrying the same load across the same distance has losses of 4.2%. For a given amount of power, a higher voltage reduces the current and thus the resistive losses in the conductor. For example, raising the voltage by a factor of 10 reduces the current by a corresponding factor of 10 and therefore the I^2R losses by a factor of 100, provided the same sized conductors are used in both cases. Even if the conductor size (cross-sectional area) is reduced 10-fold to match the lower current, the I^2R losses are still reduced 10-fold. Long-distance transmission is typically done with overhead lines at voltages of 115 to 1,200 kV. At extremely high voltages, more than 2,000 kV exists between conductor and ground, corona discharge losses are so large that they can offset the lower resistive losses in the line conductors. Measures to reduce corona losses include conductors having larger diameters; often hollow to save weight, or bundles of two or more conductors.

Factors that affect the resistance, and thus loss, of conductors used in transmission and distribution lines include temperature, spiraling, and the skin effect. The resistance of a conductor increases with its temperature. Temperature changes in electric power lines can have a significant effect on power losses in the line. Spiraling, which refers to the increase in conductor resistance due to the way stranded conductors spiral about the center, also contributes to increases in conductor resistance. The skin effect causes the effective resistance of a conductor to increase at higher alternating current frequencies.

Transmission and distribution losses in the USA were estimated at 6.6% in 1997 and 6.5% in 2007. In general, losses are estimated from the discrepancy between power produced (as reported by power plants) and power sold to the end customers; the difference between what is produced and what is consumed constitute transmission and distribution losses, assuming no utility theft occurs.

As of 1980, the longest cost-effective distance for direct-current transmission was determined to be 7,000 km (4,300 mi). For alternating current it was 4,000 km (2,500 mi), though all transmission lines in use today are substantially shorter than this.

In any alternating current transmission line, the inductance and capacitance of the conductors can be significant. Currents that flow solely in 'reaction' to these properties of the circuit, (which together with the resistance define the impedance) constitute reactive power flow, which transmits no 'real' power to the load. These reactive currents, however, are very real and cause extra heating losses in the transmission circuit. The ratio of 'real' power (transmitted to the load) to 'apparent' power (sum of 'real' and 'reactive') is the power factor. As reactive current increases, the reactive power increases and the power factor decreases. For transmission systems with low power factor, losses are higher than for systems with high power factor. Utilities add capacitor banks, reactors and other components (such as phase-shifting transformers; static VAR compensators; and flexible AC transmission systems, FACTS) throughout the system to compensate for the reactive power flow and reduce the losses in power transmission and stabilize system voltages. These measures are collectively called 'reactive support'.

Transposition

Current flowing through transmission lines induces a magnetic field that surrounds the lines of each phase and affects the inductance of the surrounding conductors of other phases. The mutual inductance of the conductors is partially dependent on the physical orientation of the lines with respect to each other. Three-phase power transmission lines are conventionally strung with phases separated on different vertical levels. The mutual inductance seen by a conductor of the phase in the middle of the other two phases will be different than the inductance seen by the conductors on the top or bottom. Because of this phenomenon, conductors must be periodically transposed along the length of the transmission line so that each phase sees equal time in each relative position to balance out the mutual inductance seen by all three phases. To accomplish this, line position is swapped at specially designed transposition towers at regular intervals along the length of the transmission line in various transposition schemes.

Subtransmission

Subtransmission is part of an electric power transmission system that runs at relatively lower voltages. It is uneconomical to connect all distribution substations to the high main transmission voltage, because the equipment is larger and more expensive. Typically, only larger substations connect with this high voltage. It is stepped down and sent to smaller substations in towns and neighborhoods. Subtransmission circuits are usually arranged in loops so that a single line failure does not cut off service to a large number of customers for more than a short time. Loops can be "normally closed", where loss of one circuit should result in no interruption, or "normally open" where substations can switch to a backup supply. While subtransmission circuits are usually carried on overhead lines, in urban areas buried cable may be used. The lower-voltage subtransmission lines use less right-of-way and simpler structures; it is much more feasible to put them underground where needed. Higher-voltage lines require more space and are usually above-ground since putting them underground is very expensive.

A 115 kV subtransmission line in the Philippines, along with 20 kV distribution lines and a street light, all mounted in a wood subtransmission pole

There is no fixed cutoff between subtransmission and transmission, or subtransmission and distribution. The voltage ranges overlap somewhat. Voltages of 69 kV, 115 kV and 138 kV are often used for subtransmission in North America. As power systems evolved, voltages formerly used for transmission were used for subtransmission, and subtransmission voltages became distribution voltages. Like transmission, subtransmission moves relatively large amounts of power, and like distribution, subtransmission covers an area instead of just point to point.

Subtransmission line tower of wood H-frame construction

Transmission Grid Exit

At the substations, transformers reduce the voltage to a lower level for distribution to commercial and residential users. This distribution is accomplished with a combination of sub-transmission (33 kV to 132 kV) and distribution (3.3 to 25 kV). Finally, at the point of use, the energy is transformed to low voltage (varying by country and customer requirements— Mains electricity by country).

Advantage of High-voltage Power Transmission

High-voltage power transmission allow for lesser resistive losses over long distances in the wiring.

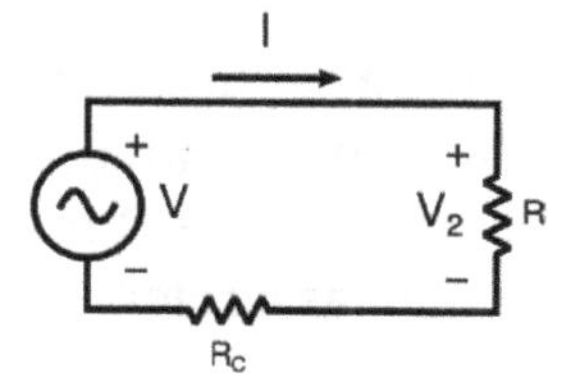

Electrical grid without a transformer.

In a very simplified model, assume the electrical grid delivers electricity from a generator (modelled as an ideal voltage source with voltage V, delivering a power P_V) to a single point of consumption, modelled by a pure resistance R, when the wires are long enough to have a significant resistance R_C.

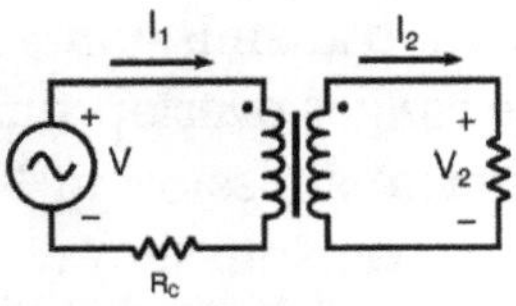

Electrical grid with a transformer.

If the resistance are simply in series without any transformer between them, the circuit acts as a voltage divider, because the same current $I = \frac{V}{R+R_C}$ runs through the wire resistance and the powered device. As a consequence, the useful power (used at the point of consumption) is:

$$P_R = V_2 \times I = \frac{V^2}{R+R_C} = \frac{R}{R+R_C} P_V$$

Assume now that a transformer converts high-voltage, low-current electricity transported by the wires into low-voltage, high-current electricity for use at the consumption point. If we suppose it is an ideal transformer with a voltage ratio of a (i.e., the voltage is divided by a and the current is multiplied by a in the secondary branch, compared to the primary branch), then the circuit is again equivalent to a voltage divider, but the transformer-consumption branch has an apparent resistance of a^2R. The useful power is then:

$$P_R = V_2 \times I_2 = \frac{V^2}{a^2R+R_C} = \frac{a^2R}{a^2R+R_C} P_V$$

For $a > 1$ (i.e. conversion of high voltage to low voltage near the consumption point), a larger fraction of the generator's power is transmitted to the consumption point and a lesser fraction is lost to Joule heating.

Modeling: The Transmission Matrix

Oftentimes, we are only interested in the terminal characteristics of the transmission line, which are the voltage and current at the sending and receiving ends. The transmission line itself is then modeled as a "black box" and a 2 by 2 transmission matrix is used to model its behavior, as follows:

The parameters A, B, C, and D differ depending on how the desired model handles the line's resistance (R), inductance (L), capacitance (C), and shunt (parallel) admittance Y. The four main models are the short line approximation, the medium line approximation, the long line approximation (with distributed parameters), and the lossless line. In all models described, a capital letter such as R refers to the total quantity summed over the line and a lowercase letter such as c refers to the per-unit-length quantity.

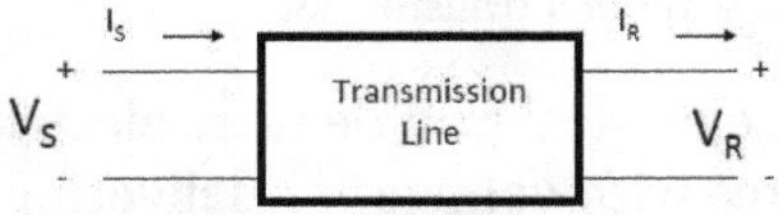

$$\begin{bmatrix} V_S \\ I_S \end{bmatrix} = \begin{bmatrix} A & B \\ C & D \end{bmatrix} \begin{bmatrix} V_R \\ I_R \end{bmatrix}$$

The line is assumed to be a reciprocal, symmetrical network, meaning that the receiving and sending labels can be switched with no consequence. The transmission matrix T also has the following properties:

- det(T) = AD - BC = 1
- A = D

The lossless line approximation, which is the least accurate model, is often used on short lines when the inductance of the line is much greater than its resistance. For this approximation, the voltage and current are identical at the sending and receiving ends.

The short line approximation is normally used for lines less than 50 miles long. For a short line, only a series impedance Z is considered, while C and Y are ignored. The final result is that A = D = 1 per unit, B = Z ohms, and C = 0.

$$\begin{bmatrix} V_S \\ I_S \end{bmatrix} = \begin{bmatrix} 1 & Z \\ 0 & 1 \end{bmatrix} \begin{bmatrix} V_R \\ I_R \end{bmatrix}$$

Transmission matrix for the short line approximation

The medium line approximation is used for lines between 50 and 150 miles long. In this model, the series impedance and the shunt admittance are considered, with half of the shunt admittance being placed at each end of the line. This circuit is often referred to as a "nominal pi" circuit because of the shape that is taken on when admittance is placed on both sides of the circuit diagram. The analysis of the medium line brings one to the following result:

$$A = D = 1 + \frac{YZ}{2} \; per\ unit$$

$$B = Z\ \Omega$$

$$C = Y\left(1 + \frac{YZ}{4}\right)\ S$$

The long line model is used when a higher degree of accuracy is needed or when the line under consideration is more than 150 miles long. Series resistance and shunt admittance are considered as distributed parameters, meaning each differential length of the line has a corresponding differential resistance and shunt admittance. The following result can be applied at any point along the transmission line, where gamma is defined as the propagation constant.

$$A = D = \cosh(\gamma x)\ per\ unit$$

$$B = Z_c \sinh(\gamma x)\ \Omega$$

$$C = \frac{1}{Z_c} \sinh(\gamma x)\ S$$

To find the voltage and current at the end of the long line, x should be replaced with L (the line length) in all parameters of the transmission matrix.

High-voltage Direct Current

High-voltage direct current (HVDC) is used to transmit large amounts of power over long distances or for interconnections between asynchronous grids. When electrical energy is to be transmitted over very long distances, the power lost in AC transmission becomes appreciable and it is less expensive to use direct current instead of alternating current. For a very long transmission line, these lower losses (and reduced construction cost of a DC line) can offset the additional cost of the required converter stations at each end.

HVDC is also used for submarine cables because AC cannot be supplied over distances of more than about 30 kilometres (19 mi), due to the fact that the cables produce too much reactive power. In these cases special high-voltage cables for DC are used. Submarine HVDC systems are often used to connect the electricity grids of islands, for example, between Great Britain and continental Europe, between Great Britain and Ireland, between Tasmania and the Australian mainland, and between the North and South Islands of New Zealand. Submarine connections up to 600 kilometres (370 mi) in length are presently in use.

HVDC links can be used to control problems in the grid with AC electricity flow. The power transmitted by an AC line increases as the phase angle between source end voltage and destination ends increases, but too large a phase angle will allow the systems at either end of the line to fall out of step. Since the power flow in a DC link is controlled independently of the phases of the AC networks at either end of the link, this phase angle limit does not exist, and a DC link is always able to transfer its full rated power. A DC link therefore stabilizes the AC grid at either end, since power flow and phase angle can then be controlled independently.

As an example, to adjust the flow of AC power on a hypothetical line between Seattle and Boston would require adjustment of the relative phase of the two regional electrical grids. This is an everyday occurrence in AC systems, but one that can become disrupted when AC system components fail and place unexpected loads on the remaining working grid system. With an HVDC line instead, such an interconnection would: (1) Convert AC in Seattle into HVDC; (2) Use HVDC for the 3,000 miles of cross-country transmission; and (3) Convert the HVDC to locally synchronized AC in Boston, (and possibly in other cooperating cities along the transmission route). Such a system could be less prone to failure if parts of it were suddenly shut down. One example of a long DC transmission line is the Pacific DC Intertie located in the Western United States.

Capacity

The amount of power that can be sent over a transmission line is limited. The origins of the limits vary depending on the length of the line. For a short line, the heating of conductors due to line losses sets a thermal limit. If too much current is drawn, conductors may sag too close to the ground, or conductors and equipment may be damaged by overheating. For intermediate-length lines on the order of 100 km (62 mi), the limit is set by the voltage drop in the line. For longer AC lines, system stability sets the limit to the power that can be transferred. Approximately, the power flowing over an AC line is proportional to the cosine of the phase angle of the voltage and current at the receiving and transmitting ends. This angle varies depending on system loading and generation. It is undesirable for the angle to approach 90 degrees, as the power flowing decreases but the resistive losses remain. Very approximately, the allowable product of line length and maximum load is proportional to the square of the system voltage. Series capacitors or phase-shifting transformers are used on long lines to improve stability. High-voltage direct current lines are restricted only by thermal and voltage drop limits, since the phase angle is not material to their operation.

Up to now, it has been almost impossible to foresee the temperature distribution along the cable route, so that the maximum applicable current load was usually set as a compromise between understanding of operation conditions and risk minimization. The availability of industrial distributed temperature sensing (DTS) systems that measure in real time temperatures all along the cable is a first step in monitoring the transmission system capacity. This monitoring solution is based on using passive optical fibers as temperature sensors, either integrated directly inside a high voltage cable or mounted externally on the cable insulation. A solution for overhead lines is also available. In this case the optical fiber is integrated into the core of a phase wire of overhead transmission lines (OPPC). The integrated Dynamic Cable Rating (DCR) or also called Real Time Thermal Rating (RTTR) solution enables not only to continuously monitor the temperature of a high voltage cable circuit in real time, but to safely utilize the existing network capacity to its maximum. Furthermore, it provides the ability to the operator to predict the behavior of the transmission system upon major changes made to its initial operating conditions.

Control

To ensure safe and predictable operation, the components of the transmission system are controlled with generators, switches, circuit breakers and loads. The voltage, power, frequency, load factor, and reliability capabilities of the transmission system are designed to provide cost effective performance for the customers.

Load Balancing

The transmission system provides for base load and peak load capability, with safety and fault tolerance margins. The peak load times vary by region largely due to the industry mix. In very hot and very cold climates home air conditioning and heating loads have an effect on the overall load. They are typically highest in the late afternoon in the hottest part of the year and in mid-mornings and mid-evenings in the coldest part of the year. This makes the power requirements vary by the season and the time of day.

Distribution system designs always take the base load and the peak load into consideration.

The transmission system usually does not have a large buffering capability to match the loads with the generation. Thus generation has to be kept matched to the load, to prevent overloading failures of the generation equipment.

Multiple sources and loads can be connected to the transmission system and they must be controlled to provide orderly transfer of power. In centralized power generation, only local control of generation is necessary, and it involves synchronization of the generation units, to prevent large transients and overload conditions.

In distributed power generation the generators are geographically distributed and the process to bring them online and offline must be carefully controlled. The load control signals can either be sent on separate lines or on the power lines themselves. Voltage and frequency can be used as signalling mechanisms to balance the loads.

In voltage signaling, the variation of voltage is used to increase generation. The power added by any system increases as the line voltage decreases. This arrangement is stable in principle. Voltage-based regulation is complex to use in mesh networks, since the individual components and setpoints would need to be reconfigured every time a new generator is added to the mesh.

In frequency signaling, the generating units match the frequency of the power transmission system. In droop speed control, if the frequency decreases, the power is increased. (The drop in line frequency is an indication that the increased load is causing the generators to slow down.)

Wind turbines, vehicle-to-grid and other distributed storage and generation systems can be connected to the power grid, and interact with it to improve system operation.

Failure Protection

Under excess load conditions, the system can be designed to fail gracefully rather than all at once. Brownouts occur when the supply power drops below the demand. Blackouts occur when the supply fails completely.

Rolling blackouts (also called load shedding) are intentionally engineered electrical power outages, used to distribute insufficient power when the demand for electricity exceeds the supply.

Communications

Operators of long transmission lines require reliable communications for control of the power grid and, often, associated generation and distribution facilities. Fault-sensing protective relays at each end of the line must communicate to monitor the flow of power into and out of the protected line section so that faulted conductors or equipment can be quickly de-energized and the balance of the system restored. Protection of the transmission line from short circuits and other faults is usually so critical that common carrier telecommunications are insufficiently reliable, and in remote areas a common carrier may not be available. Communication systems associated with a transmission project may use:

- Microwaves
- Power line communication
- Optical fibers

Rarely, and for short distances, a utility will use pilot-wires strung along the transmission line path. Leased circuits from common carriers are not preferred since availability is not under control of the electric power transmission organization.

Transmission lines can also be used to carry data: this is called power-line carrier, or PLC. PLC signals can be easily received with a radio for the long wave range.

Optical fibers can be included in the stranded conductors of a transmission line, in the overhead shield wires. These cables are known as optical ground wire (OPGW). Sometimes a standalone cable is used, all-dielectric self-supporting (ADSS) cable, attached to the transmission line cross arms.

Some jurisdictions, such as Minnesota, prohibit energy transmission companies from selling surplus communication bandwidth or acting as a telecommunications common carrier. Where the regulatory structure permits, the utility can sell capacity in extra dark fibers to a common carrier, providing another revenue stream.

Electricity Market Reform

Some regulators regard electric transmission to be a natural monopoly and there are moves in many countries to separately regulate transmission.

Spain was the first country to establish a regional transmission organization. In that country, transmission operations and market operations are controlled by separate companies. The transmission system operator is Red Eléctrica de España (REE) and the wholesale electricity market operator is Operador del Mercado Ibérico de Energía – Polo Español, S.A. (OMEL) . Spain's transmission system is interconnected with those of France, Portugal, and Morocco.

In the United States and parts of Canada, electrical transmission companies operate independently of generation and distribution companies.

Cost of Electric Power Transmission

The cost of high voltage electricity transmission (as opposed to the costs of electric power distribution) is comparatively low, compared to all other costs arising in a consumer's electricity bill. In the UK, transmission costs are about 0.2p/kWh compared to a delivered domestic price of around 10p/kWh.

Research evaluates the level of capital expenditure in the electric power T&D equipment market will be worth $128.9bn in 2011.

Merchant Transmission

Merchant transmission is an arrangement where a third party constructs and operates electric transmission lines through the franchise area of an unrelated utility.

Operating merchant transmission projects in the United States include the Cross Sound Cable from Shoreham, New York to New Haven, Connecticut, Neptune RTS Transmission Line from Sayreville, N.J., to Newbridge, N.Y, and Path 15 in California. Additional projects are in development or have been proposed throughout the United States, including the Lake Erie Connector, an underwater transmission line proposed by ITC Holdings Corp., connecting Ontario to load serving entities in the PJM Interconnection region.

There is only one unregulated or market interconnector in Australia: Basslink between Tasmania and Victoria. Two DC links originally implemented as market interconnectors, Directlink and Murraylink, have been converted to regulated interconnectors. NEMMCO

A major barrier to wider adoption of merchant transmission is the difficulty in identifying who benefits from the facility so that the beneficiaries will pay the toll. Also, it is difficult for a merchant transmission line to compete when the alternative transmission lines are subsidized by other utility businesses.

Health concerns

Some large studies, including a large study in the United States, have failed to find any link between living near power lines and developing any sickness or diseases, such as cancer. A 1997 study found that it did not matter how close one was to a power line or a sub-station, there was no increased risk of cancer or illness.

The mainstream scientific evidence suggests that low-power, low-frequency, electromagnetic radiation associated with household currents and high transmission power lines does not constitute a short or long term health hazard. Some studies, however, have found statistical correlations between various diseases and living or working near power lines. No adverse health effects have been substantiated for people not living close to powerlines.

There are established biological effects for acute high level exposure to magnetic fields well above 100 µT (1 G). In a residential setting, there is "limited evidence of carcinogenicity in humans and less than sufficient evidence for carcinogenicity in experimental animals", in particular, childhood leukemia, associated with average exposure to residential power-frequency magnetic field above 0.3 μT (3 mG) to 0.4 μT (4 mG). These levels exceed average residential power-frequency magnetic fields in homes, which are about 0.07 μT (0.7 mG) in Europe and 0.11 μT (1.1 mG) in North America.

The Earth's natural geomagnetic field strength varies over the surface of the planet between 0.035 mT - 0.07 mT (35 μT - 70 μT or 0.35 G - 0.7 G) while the International Standard for the continuous exposure limit is set at 40 mT (40,000 μT or 400 G) for the general public.

Tree Growth Regulator and Herbicide Control Methods may be used in transmission line right of ways which may have health effects.

United States Government Policy

Historically, local governments have exercised authority over the grid and have significant disincentives to encourage actions that would benefit states other than their own. Localities with cheap electricity have a disincentive to encourage making interstate commerce in electricity trading easier, since other regions will be able to compete for local energy and drive up rates. For example, some regulators in Maine do not wish to address congestion problems because the congestion serves to keep Maine rates low. Further, vocal local constituencies can block or slow permitting by pointing to visual impact, environmental, and perceived health concerns. In the US, generation is growing four times faster than transmission, but big transmission upgrades require the coordination of multiple states, a multitude of interlocking permits, and cooperation between a significant portion of the 500 companies that own the grid. From a policy perspective, the control of the grid is balkanized, and even former energy secretary Bill Richardson refers to it as a third world grid. There have been efforts in the EU and US to confront the problem. The US national security interest in significantly growing transmission capacity drove passage of the 2005 energy act giving the Department of Energy the authority to approve transmission if states refuse to act. However, soon after the Department of Energy used its power to designate two National Interest Electric Transmission Corridors, 14 senators signed a letter stating the DOE was being too aggressive.

Special Transmission

Grids for Railways

In some countries where electric locomotives or electric multiple units run on low frequency AC power, there are separate single phase traction power networks operated by the railways. Prime examples are countries in Europe (including Austria, Germany and Switzerland) which utilize the older AC technology based on 16 2/3 Hz (Norway and Sweden also use this frequency but use conversion from the 50Hz public supply; Sweden has a 16 2/3 Hz traction grid but only for part of the system).

Superconducting Cables

High-temperature superconductors (HTS) promise to revolutionize power distribution by providing lossless transmission of electrical power. The development of superconductors with transition temperatures higher than the boiling point of liquid nitrogen has made the concept of superconducting power lines commercially feasible, at least for high-load applications. It has been estimated that the waste would be halved using this method, since the necessary refrigeration equipment would consume about half the power saved by the elimination of the majority of resistive losses. Some companies such as Consolidated Edison and American Superconductor have already begun commercial production of such systems. In one hypothetical future system called a SuperGrid, the cost of cooling would be eliminated by coupling the transmission line with a liquid hydrogen pipeline.

Superconducting cables are particularly suited to high load density areas such as the business district of large cities, where purchase of an easement for cables would be very costly.

HTS transmission lines				
Location	**Length (km)**	**Voltage (kV)**	**Capacity (GW)**	**Date**
Carrollton, Georgia				2000
Albany, New York	0.35	34.5	0.048	2006
Long Island	0.6	130	0.574	2008
Tres Amigas			5	Proposed 2013
Manhattan: Project Hydra				Proposed 2014
Essen, Germany	1	10	0.04	2014

Single wire Earth Return

Single-wire earth return (SWER) or single wire ground return is a single-wire transmission line for supplying single-phase electrical power for an electrical grid to remote areas at low cost. It is principally used for rural electrification, but also finds use for larger isolated loads such as water pumps. Single wire earth return is also used for HVDC over submarine power cables.

Wireless Power Transmission

Both Nikola Tesla and Hidetsugu Yagi attempted to devise systems for large scale wireless power transmission in the late 1800s and early 1900s, with no commercial success.

In November 2009, LaserMotive won the NASA 2009 Power Beaming Challenge by powering a cable climber 1 km vertically using a ground-based laser transmitter. The system produced up to 1 kW of power at the receiver end. In August 2010, NASA contracted with private companies to pursue the design of laser power beaming systems to power low earth orbit satellites and to launch rockets using laser power beams.

Wireless power transmission has been studied for transmission of power from solar power satellites to the earth. A high power array of microwave or laser transmitters would beam power to a rectenna. Major engineering and economic challenges face any solar power satellite project.

Security of Control Systems

The Federal government of the United States admits that the power grid is susceptible to cyber-warfare. The United States Department of Homeland Security works with industry to identify vulnerabilities and to help industry enhance the security of control system networks, the federal government is also working to ensure that security is built in as the U.S. develops the next generation of 'smart grid' networks.

Power-system Protection

Power-system protection is a branch of electrical power engineering that deals with the protection of electrical power systems from faults through the isolation of faulted parts from the rest of the electrical network. The objective of a protection scheme is to keep the power system stable by isolating only the components that are under fault, whilst leaving as much of the network as possible still in operation. Thus, protection schemes must apply with very pragmatic and pessimistic approach to clearing system faults. The devices that are used to protect the power systems from faults are called protection devices.

Components

Protection systems usually comprise five components:

- Current and voltage transformers to step down the high voltages and currents of the electrical power system to convenient levels for the relays to deal with
- Protective relays to sense the fault and initiate a trip, or disconnection, order;
- Circuit breakers to open/close the system based on relay and autorecloser commands;
- Batteries to provide power in case of power disconnection in the system.

- Communication channels to allow analysis of current and voltage at remote terminals of a line and to allow remote tripping of equipment.

For parts of a distribution system, fuses are capable of both sensing and disconnecting faults.

Failures may occur in each part, such as insulation failure, fallen or broken transmission lines, incorrect operation of circuit breakers, short circuits and open circuits. Protection devices are installed with the aims of protection of assets, and ensure continued supply of energy.

Switchgear is a combination of electrical disconnect switches, fuses or circuit breakers used to control, protect and isolate electrical equipment. Switches are safe to open under normal load current, while protective devices are safe to open under fault current.

Protective Device

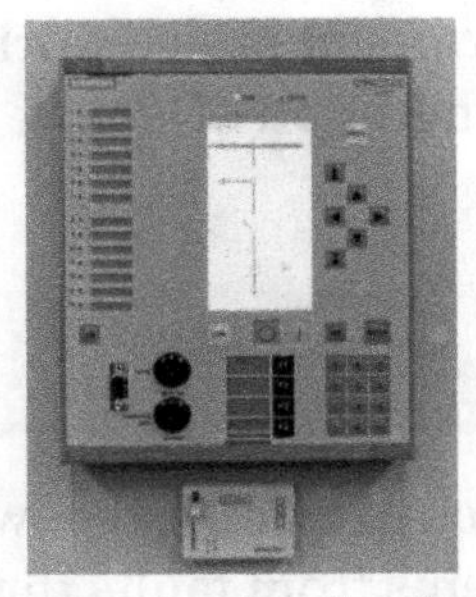

A digital (numeric) multifunction protective relay for distribution networks. A single such device can replace many single-function electromechanical relays, and provides self-testing and communication functions.

- Protective relays control the tripping of the circuit breakers surrounding the faulted part of the network
- Automatic operation, such as auto-re-closing or system restart
- Monitoring equipment which collects data on the system for post event analysis

While the operating quality of these devices, and especially of protective relays, is always critical, different strategies are considered for protecting the different parts of the system. Very important equipment may have completely redundant and independent protective systems, while a minor branch distribution line may have very simple low-cost protection.

There are three parts of protective devices:

- Instrument transformer: current or potential (CT or VT)
- Relay

- Circuit breaker

Advantages of protected devices with these three basic components include safety, economy, and accuracy.

- Safety: Instrument transformers create electrical isolation from the power system, and thus establishing a safer environment for personnel working with the relays.
- Economy: Relays are able to be simpler, smaller, and cheaper given lower-level relay inputs.
- Accuracy: Power system voltages and currents are accurately reproduced by instrument transformers over large operating ranges.

Types of Protection

- Generator sets – In a power plant, the protective relays are intended to prevent damage to alternators or to the transformers in case of abnormal conditions of operation, due to internal failures, as well as insulating failures or regulation malfunctions. Such failures are unusual, so the protective relays have to operate very rarely. If a protective relay fails to detect a fault, the resulting damage to the alternator or to the transformer might require costly equipment repairs or replacement, as well as income loss from the inability to produce and sell energy.
- High-voltage transmission network – Protection on the transmission and distribution serves two functions: Protection of plant and protection of the public (including employees). At a basic level, protection looks to disconnect equipment which experience an overload or a short to earth. Some items in substations such as transformers might require additional protection based on temperature or gas pressure, among others.
- Overload and back-up for distance (overcurrent) – Overload protection requires a current transformer which simply measures the current in a circuit. There are two types of overload protection: instantaneous overcurrent and time overcurrent (TOC). Instantaneous overcurrent requires that the current exceeds a predetermined level for the circuit breaker to operate. TOC protection operates based on a current vs time curve. Based on this curve if the measured current exceeds a given level for the preset amount of time, the circuit breaker or fuse will operate.
- Earth fault ("ground fault" in the United States) – Earth fault protection again requires current transformers and senses an imbalance in a three-phase circuit. Normally the three phase currents are in balance, i.e. roughly equal in magnitude. If one or two phases become connected to earth via a low impedance path,

their magnitudes will increase dramatically, as will current imbalance. If this imbalance exceeds a pre-determined value, a circuit breaker should operate. Restricted earth fault protection is a type of earth fault protection which looks for earth fault between two sets current transformers (hence restricted to that zone).

- Distance (impedance relay) – Distance protection detects both voltage and current. A fault on a circuit will generally create a sag in the voltage level. If the ratio of voltage to current measured at the relay terminals, which equates to an impedance, lands within a predetermined level the circuit breaker will operate. This is useful for reasonable length lines, lines longer than 10 miles, because its operating characteristics are based on the line characteristics. This means that when a fault appears on the line the impedance setting in the relay is compared to the apparent impedance of the line from the relay terminals to the fault. If the relay setting is determined to be below the apparent impedance it is determined that the fault is within the zone of protection. When the transmission line length is too short, less than 10 miles, distance protection becomes more difficult to coordinate. In these instances the best choice of protection is current differential protection.
- Back-up – The objective of protection is to remove only the affected portion of plant and nothing else. A circuit breaker or protection relay may fail to operate. In important systems, a failure of primary protection will usually result in the operation of back-up protection. Remote back-up protection will generally remove both the affected and unaffected items of plant to clear the fault. Local back-up protection will remove the affected items of the plant to clear the fault.
- Low-voltage networks – The low-voltage network generally relies upon fuses or low-voltage circuit breakers to remove both overload and earth faults.

Coordination

Protective device coordination is the process of determining the "best fit" timing of current interruption when abnormal electrical conditions occur. The goal is to minimize an outage to the greatest extent possible. Historically, protective device coordination was done on translucent log–log paper. Modern methods normally include detailed computer based analysis and reporting.

Protection coordination is also handled through dividing the power system into protective zones. If a fault were to occur in a given zone, necessary actions will be executed to isolate that zone from the entire system. Zone definitions account for generators, buses, transformers, transmission and distribution lines, and motors. Additionally, zones possess the following features: zones overlap, overlap regions denote circuit breakers, and all circuit breakers in a given zone with a fault will open in order to isolate the fault.

Overlapped regions are created by two sets of instrument transformers and relays for each circuit breaker. They are designed for redundancy to eliminate unprotected areas; however, overlapped regions are devised to remain as small as possible such that when a fault occurs in an overlap region and the two zones which encompass the fault are isolated, the sector of the power system which is lost from service is still small despite two zones being isolated.

Disturbance-monitoring Equipment

Disturbance-monitoring equipment (DME) monitors and records system data pertaining to a fault. DME accomplish three main purposes:

1. model validation,
2. disturbance investigation, and
3. assessment of system protection performance.

DME devices include:

- Sequence of event recorders, which record equipment response to the event
- Fault recorders, which record actual waveform data of the system primary voltages and currents.
- Dynamic disturbance recorders (DDRs), which record incidents that portray power system behavior during dynamic events such as low frequency (0.1 Hz – 3 Hz) oscillations and abnormal frequency or voltage excursions

Performance Measures

Protection engineers define dependability as the tendency of the protection system to operate correctly for in-zone faults. They define security as the tendency not to operate for out-of-zone faults. Both dependability and security are reliability issues. Fault tree analysis is one tool with which a protection engineer can compare the relative reliability of proposed protection schemes. Quantifying protection reliability is important for making the best decisions on improving a protection system, managing dependability versus security tradeoffs, and getting the best results for the least money. A quantitative understanding is essential in the competitive utility industry.

Performance and design criteria for system-protection devices include reliability, selectivity, speed, cost, and simplicity.

- Reliability: Devices must function consistently when fault conditions occur, regardless of possibly being idle for months or years. Without this reliability, systems may result in high costly damages.

- Selectivity: Devices must avoid unwarranted, false trips.
- Speed: Devices must function quickly to reduce equipment damage and fault duration, with only very precise intentional time delays.
- Economy: Devices must provide maximum protection at minimum cost.
- Simplicity: Devices must minimize protection circuitry and equipment.

Transformer

Pole-mounted distribution transformer with center-tapped secondary winding used to provide 'split-phase' power for residential and light commercial service, which in North America is typically rated 120/240 V.

A transformer is an electrical device that transfers electrical energy between two or more circuits through electromagnetic induction. Electromagnetic induction produces an electromotive force within a conductor which is exposed to time varying magnetic fields. Transformers are used to increase or decrease the alternating voltages in electric power applications.

A varying current in the transformer's primary winding creates a varying magnetic flux in the transformer core and a varying field impinging on the transformer's secondary winding. This varying magnetic field at the secondary winding induces a varying electromotive force (EMF) or voltage in the secondary winding due to electromagnetic induction. Making use of Faraday's Law (discovered in 1831) in conjunction with high magnetic permeability core properties, transformers can be designed to efficiently change AC voltages from one voltage level to another within power networks.

Since the invention of the first constant potential transformer in 1885, transformers have become essential for the transmission, distribution, and utilization of alternating current electrical energy. A wide range of transformer designs is encountered in electronic and electric power applications. Transformers range in size from RF transformers less than a cubic centimeter in volume to units interconnecting the power grid weighing hundreds of tons.

Basic Principles

Ideal Transformer

Ideal transformer equations (eq.)

By Faraday's law of induction:

$$V_S = -N_S \frac{d\Phi}{dt} \ldots (1)$$

$$V_P = -N_P \frac{d\Phi}{dt} \ldots (2)$$

Combining ratio of (1) & (2)

Turns ratio $= \frac{V_P}{V_S} = \frac{-N_P}{-N_S} = a \ldots (3)$ where

for step-down transformers, a > 1

for step-up transformers, a < 1

By law of conservation of energy, apparent, real and reactive power are each conserved in the input and output

$$S = I_P V_P = I_S V_S \ldots (4)$$

Combining (3) & (4) with this endnote yields the ideal transformer identity

$$\frac{V_P}{V_S} = \frac{I_S}{I_P} = \frac{N_P}{N_S} = \sqrt{\frac{L_P}{L_S}} = a \,. (5)$$

By Ohm's law and ideal transformer identity

$$Z_L = \frac{V_S}{I_S} \ldots (6)$$

Apparent load impedance Z'_L (Z_L referred to the primary)

$$Z'_L = \frac{V_P}{I_P} = \frac{aV_S}{I_S / a} = a^2 \frac{V_S}{I_S} = a^2 Z_L \,. (7)$$

For simplification or approximation purposes, it is very common to analyze the transformer as an ideal transformer model as presented in the two images. An ideal transformer is a theoretical, linear transformer that is lossless and perfectly coupled; that is, there are no energy losses and flux is completely confined within the magnetic core.

Perfect coupling implies infinitely high core magnetic permeability and winding inductances and zero net magnetomotive force.

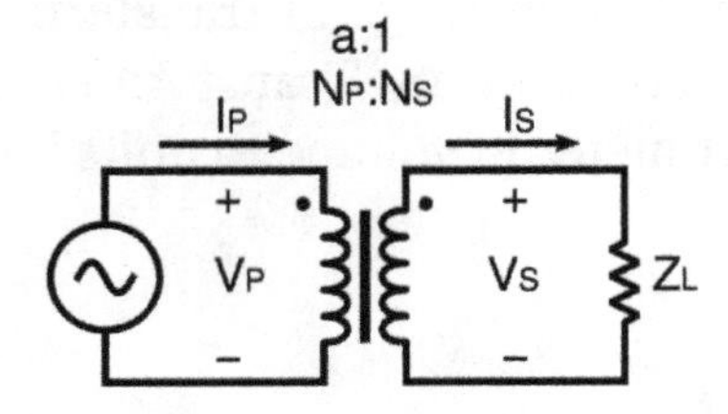

Ideal transformer connected with source V_P on primary and load impedance Z_L on secondary, where $0 < Z_L < \infty$.

A varying current in the transformer's primary winding creates a varying magnetic flux in the core and a varying magnetic field impinging on the secondary winding. This varying magnetic field at the secondary induces a varying electromotive force (EMF) or voltage in the secondary winding. The primary and secondary windings are wrapped around a core of infinitely high magnetic permeability so that all of the magnetic flux passes through both the primary and secondary windings. With a voltage source connected to the primary winding and load impedance connected to the secondary winding, the transformer currents flow in the indicated directions.

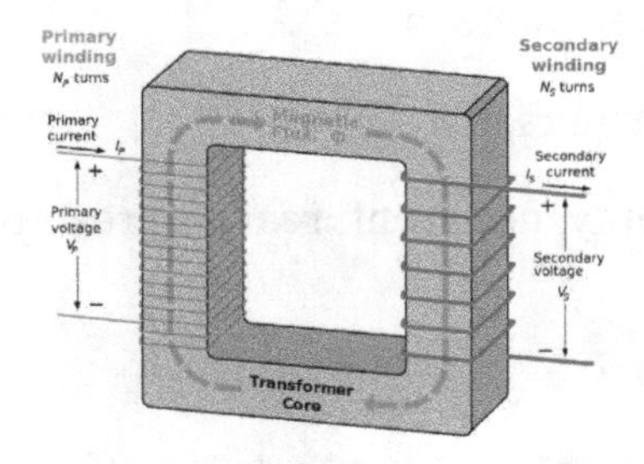

Ideal transformer and induction law

According to Faraday's Law, since the same magnetic flux passes through both the primary and secondary windings in an ideal transformer, a voltage is induced in each winding, according to eq. (1) in the secondary winding case, according to eq. (2) in the primary winding case. The primary EMF is sometimes termed counter EMF. This is in accordance with Lenz's law, which states that induction of EMF always opposes development of any such change in magnetic field.

The transformer winding voltage ratio is thus shown to be directly proportional to the winding turns ratio according to eq. (3).

According to the law of conservation of energy, any load impedance connected to the ideal transformer's secondary winding results in conservation of apparent, real and reactive power consistent with eq. (4).

The ideal transformer identity shown in eq. (5) is a reasonable approximation for the typical commercial transformer, with voltage ratio and winding turns ratio both being inversely proportional to the corresponding current ratio.

Instrument transformer, with polarity dot and X1 markings on LV side terminal

By Ohm's law and the ideal transformer identity:

- the secondary circuit load impedance can be expressed as eq. (6)
- the apparent load impedance referred to the primary circuit is derived in eq. (7) to be equal to the turns ratio squared times the secondary circuit load impedance.

Polarity

A dot convention is often used in transformer circuit diagrams, nameplates or terminal markings to define the relative polarity of transformer windings. Positively increasing instantaneous current entering the primary winding's dot end induces positive polarity voltage at the secondary winding's dot end.

Real Transformer

Deviations from ideal

The ideal transformer model neglects the following basic linear aspects in real transformers.

Core losses, collectively called magnetizing current losses, consist of

- Hysteresis losses due to nonlinear application of the voltage applied in the transformer core, and
- Eddy current losses due to joule heating in the core that are proportional to the square of the transformer's applied voltage.

Whereas windings in the ideal model have no resistances and infinite inductances, the windings in a real transformer have finite non-zero resistances and inductances associated with:

- Joule losses due to resistance in the primary and secondary windings
- Leakage flux that escapes from the core and passes through one winding only resulting in primary and secondary reactive impedance.

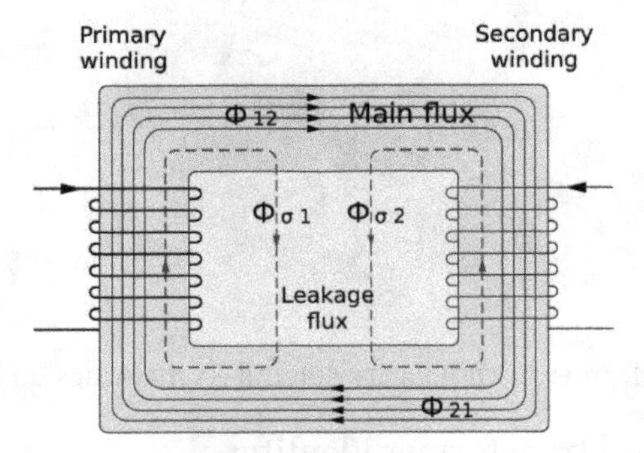

Leakage flux of a transformer

Leakage Flux

The ideal transformer model assumes that all flux generated by the primary winding links all the turns of every winding, including itself. In practice, some flux traverses paths that take it outside the windings. Such flux is termed leakage flux, and results in leakage inductance in series with the mutually coupled transformer windings. Leakage flux results in energy being alternately stored in and discharged from the magnetic fields with each cycle of the power supply. It is not directly a power loss, but results in inferior voltage regulation, causing the secondary voltage not to be directly proportional to the primary voltage, particularly under heavy load. Transformers are therefore normally designed to have very low leakage inductance.

In some applications increased leakage is desired, and long magnetic paths, air gaps, or magnetic bypass shunts may deliberately be introduced in a transformer design to limit the short-circuit current it will supply. Leaky transformers may be used to supply loads that exhibit negative resistance, such as electric arcs, mercury vapor lamps, and neon signs or for safely handling loads that become periodically short-circuited such as electric arc welders.

Air gaps are also used to keep a transformer from saturating, especially audio-frequency transformers in circuits that have a DC component flowing in the windings.

Knowledge of leakage inductance is also useful when transformers are operated in parallel. It can be shown that if the percent impedance and associated winding leakage reactance-to-resistance (X/R) ratio of two transformers were hypothetically exactly the same, the transformers would share power in proportion to their respective volt-ampere ratings (e.g. 500 kVA unit in parallel with 1,000 kVA unit, the larger unit would carry twice the current). However, the impedance tolerances of commercial transformers are significant. Also, the Z impedance and X/R ratio of different capacity transformers tends to vary, corresponding 1,000 kVA and 500 kVA units' values being, to illustrate, respectively, Z ≈ 5.75%, X/R ≈ 3.75 and Z ≈ 5%, X/R ≈ 4.75.

Equivalent Circuit

Referring to the diagram, a practical transformer's physical behavior may be represented by an equivalent circuit model, which can incorporate an ideal transformer.

Winding joule losses and leakage reactances are represented by the following series loop impedances of the model:

- Primary winding: R_P, X_P
- Secondary winding: R_S, X_S.

In normal course of circuit equivalence transformation, R_S and X_S are in practice usually referred to the primary side by multiplying these impedances by the turns ratio squared, $(N_P/N_S)^2 = a^2$.

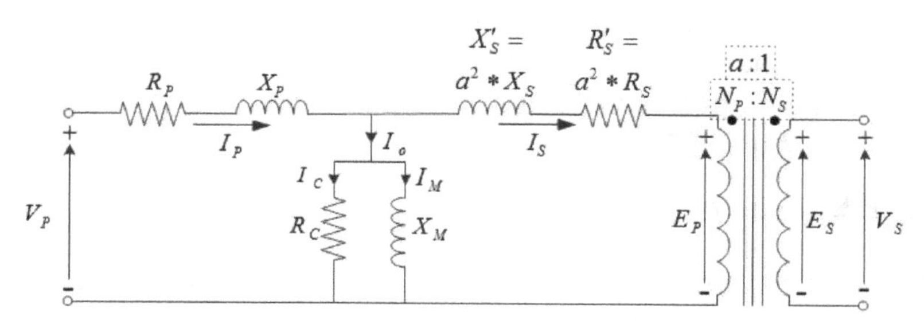

Real transformer equivalent circuit

Core loss and reactance is represented by the following shunt leg impedances of the model:

- Core or iron losses: R_C
- Magnetizing reactance: X_M.

R_C and X_M are collectively termed the magnetizing branch of the model.

Core losses are caused mostly by hysteresis and eddy current effects in the core and are proportional to the square of the core flux for operation at a given frequency. The finite permeability core requires a magnetizing current I_M to maintain mutual flux in the core. Magnetizing current is in phase with the flux, the relationship between the two being non-linear due to saturation effects. However, all impedances of the equivalent circuit shown are by definition linear and such non-linearity effects are not typically reflected in transformer equivalent circuits. With sinusoidal supply, core flux lags the induced EMF by 90°. With open-circuited secondary winding, magnetizing branch current I_o equals transformer no-load current.

The resulting model, though sometimes termed 'exact' equivalent circuit based on linearity assumptions, retains a number of approximations. Analysis may be simplified by assuming that magnetizing branch impedance is relatively high and relocating the branch to the left of the primary impedances. This introduces error but allows combination of primary and referred secondary resistances and reactances by simple summation as two series impedances.

Transformer equivalent circuit impedance and transformer ratio parameters can be derived from the following tests: open-circuit test, short-circuit test, winding resistance test, and transformer ratio test.

Basic Transformer Parameters and Construction

Effect of Frequency

Transformer Universal Emf Equation

If the flux in the core is purely sinusoidal, the relationship for either winding between its rms voltage E_{rms} of the winding, and the supply frequency f, number of turns N, core cross-sectional area a in m² and peak magnetic flux density B_{peak} in Wb/m² or T (tesla) is given by the universal EMF equation:

$$E_{rms} = \frac{2\pi fNaB_{peak}}{\sqrt{2}} \approx 4.44 fNaB_{peak}$$

If the flux does not contain even harmonics the following equation can be used for half-cycle average voltage E_{avg} of any waveshape:

By Faraday's Law of induction shown in eq. (1) and (2), transformer EMFs vary according to the derivative of flux with respect to time. The ideal transformer's core behaves linearly with time for any non-zero frequency. Flux in a real transformer's core behaves non-linearly in relation to magnetization current as the instantaneous flux increases beyond a finite linear range resulting in magnetic saturation associated with increasingly large magnetizing current, which eventually leads to transformer overheating.

The EMF of a transformer at a given flux density increases with frequency. By operating at higher frequencies, transformers can be physically more compact because a given core is able to transfer more power without reaching saturation and fewer turns are needed to achieve the same impedance. However, properties such as core loss and conductor skin effect also increase with frequency. Aircraft and military equipment employ 400 Hz power supplies which reduce core and winding weight. Conversely, frequencies used for some railway electrification systems were much lower (e.g. 16.7 Hz and 25 Hz) than normal utility frequencies (50–60 Hz) for historical reasons concerned mainly with the limitations of early electric traction motors. Consequently, the transformers used to step-down the high overhead line voltages (e.g. 15 kV) were much larger and heavier for the same power rating than those required for the higher frequencies.

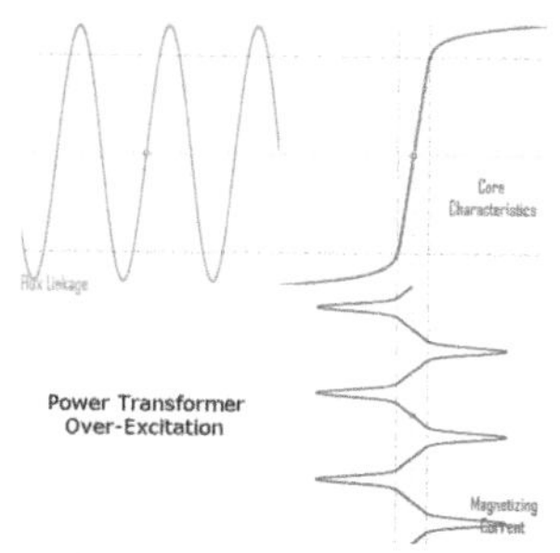

Power transformer over-excitation condition caused by decreased frequency; flux (green), iron core's magnetic characteristics (red) and magnetizing current (blue).

Operation of a transformer at its designed voltage but at a higher frequency than intended will lead to reduced magnetizing current. At a lower frequency, the magnetizing current will increase. Operation of a transformer at other than its design frequency may require assessment of voltages, losses, and cooling to establish if safe operation is practical. For example, transformers may need to be equipped with 'volts per hertz' over-excitation relays to protect the transformer from overvoltage at higher than rated frequency.

One example is in traction transformers used for electric multiple unit and high-speed train service operating across regions with different electrical standards. The converter equipment and traction transformers have to accommodate different input frequencies and voltage (ranging from as high as 50 Hz down to 16.7 Hz and rated up to 25 kV) while being suitable for multiple AC asynchronous motor and DC converters and motors with varying harmonics mitigation filtering requirements.

Large power transformers are vulnerable to insulation failure due to transient voltages with high-frequency components, such as caused in switching or by lightning.

At much higher frequencies the transformer core size required drops dramatically: a physically small and cheap transformer can handle power levels that would require a massive iron core at mains frequency. The development of switching power semiconductor devices and complex integrated circuits made switch-mode power supplies viable, to generate a high frequency from a much lower one (or DC), change the voltage level with a small transformer, and, if necessary, rectify the changed voltage.

Energy Losses

Real transformer energy losses are dominated by winding resistance joule and core losses. Transformers' efficiency tends to improve with increasing transformer capacity. The efficiency of typical distribution transformers is between about 98 and 99 percent.

As transformer losses vary with load, it is often useful to express these losses in terms of no-load loss, full-load loss, half-load loss, and so on. Hysteresis and eddy current losses are constant at all load levels and dominate overwhelmingly without load, while variable winding joule losses dominating increasingly as load increases. The no-load loss can be significant, so that even an idle transformer constitutes a drain on the electrical supply. Designing energy efficient transformers for lower loss requires a larger core, good-quality silicon steel, or even amorphous steel for the core and thicker wire, increasing initial cost. The choice of construction represents a trade-off between initial cost and operating cost.

Transformer losses arise from:

Winding Joule Losses

Current flowing through a winding's conductor causes joule heating. As frequency increases, skin effect and proximity effect causes the winding's resistance and, hence, losses to increase.

Core losses

Hysteresis Losses

Each time the magnetic field is reversed, a small amount of energy is lost due to hysteresis within the core. According to Steinmetz's formula, the heat energy due to hysteresis is given by

$\backslash W_h \approx \eta \beta_{max}^{1.6}$, and,

hysteresis loss is thus given by

$P_h \approx W_h f \approx \eta f \beta_{max}^{1.6}$

where, f is the frequency, η is the hysteresis coefficient and β_{max} is the maximum flux density, the empirical exponent of which varies from about 1.4 to 1.8 but is often given as 1.6 for iron.

Eddy Current Losses

Ferromagnetic materials are also good conductors and a core made from such a material also constitutes a single short-circuited turn throughout its entire length. Eddy currents therefore circulate within the core in a plane normal to the flux, and are responsible for resistive heating of the core material. The eddy current loss is a complex function of the square of supply frequency and inverse square of the material thickness. Eddy current losses can be reduced by making the core of a stack of plates electrically insulated from each other, rather than a solid block; all transformers operating at low frequencies use laminated or similar cores.

Magnetostriction Related Transformer Hum

Magnetic flux in a ferromagnetic material, such as the core, causes it to physically expand and contract slightly with each cycle of the magnetic field, an effect known as magnetostriction, the frictional energy of which produces an audible noise known as mains hum or transformer hum. This transformer hum is especially objectionable in transformers supplied at power frequencies and in high-frequency flyback transformers associated with television CRTs.

Stray Losses

Leakage inductance is by itself largely lossless, since energy supplied to its magnetic fields is returned to the supply with the next half-cycle. However, any leakage flux that intercepts nearby conductive materials such as the transformer's support structure will give rise to eddy currents and be converted to heat. There are also radiative losses due to the oscillating magnetic field but these are usually small.

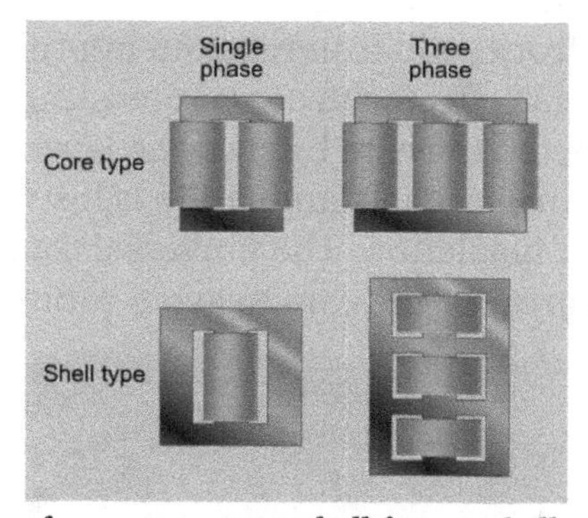

Core form = core type; shell form = shell type

Mechanical Vibration and Audible Noise Transmission

In addition to magnetostriction, the alternating magnetic field causes fluctuating forces between the primary and secondary windings. This energy incites vibration transmission in interconnected metalwork, thus amplifying audible transformer hum.

Core form and Shell form Transformers

Closed-core transformers are constructed in 'core form' or 'shell form'. When windings surround the core, the transformer is core form; when windings are surrounded by the core, the transformer is shell form. Shell form design may be more prevalent than core form design for distribution transformer applications due to the relative ease in stacking the core around winding coils. Core form design tends to, as a general rule, be more economical, and therefore more prevalent, than shell form design for high voltage power transformer applications at the lower end of their voltage and power rating ranges (less than or equal to, nominally, 230 kV or 75 MVA). At higher voltage and power ratings, shell form transformers tend to be more prevalent. Shell form design tends to be preferred for extra-high voltage and higher MVA applications because, though more labor-intensive to manufacture, shell form transformers are characterized as having inherently better kVA-to-weight ratio, better short-circuit strength characteristics and higher immunity to transit damage.

Construction

Cores

Laminated Steel Cores

Transformers for use at power or audio frequencies typically have cores made of high permeability silicon steel. The steel has a permeability many times that of free space and the core thus serves to greatly reduce the magnetizing current and confine the flux to a path which closely couples the windings. Early transformer developers soon realized that cores constructed from solid iron resulted in prohibitive eddy current losses, and their designs mitigated this effect with cores consisting of bundles of insulated iron wires. Later designs constructed the core by stacking layers of thin steel laminations, a principle that has remained in use. Each lamination is insulated from its neighbors by a thin non-conducting layer of insulation. The universal transformer equation indicates a minimum cross-sectional area for the core to avoid saturation.

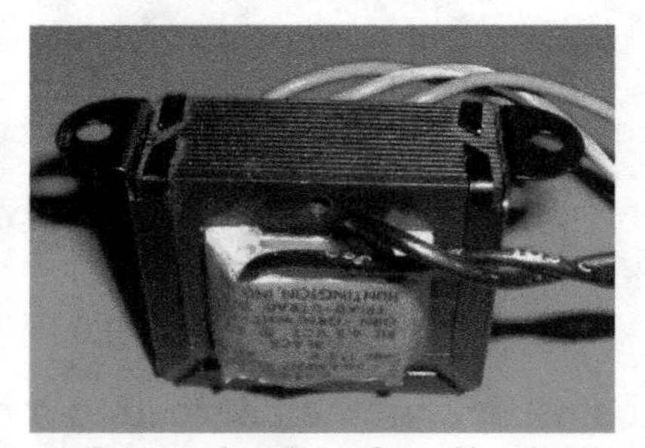

Laminated core transformer showing edge of laminations at top of photo

The effect of laminations is to confine eddy currents to highly elliptical paths that enclose little flux, and so reduce their magnitude. Thinner laminations reduce losses, but are more laborious and expensive to construct. Thin laminations are generally used on high-frequency transformers, with some of very thin steel laminations able to operate up to 10 kHz.

One common design of laminated core is made from interleaved stacks of E-shaped steel sheets capped with I-shaped pieces, leading to its name of 'E-I transformer'. Such a design tends to exhibit more losses, but is very economical to manufacture. The cut-core or C-core type is made by winding a steel strip around a rectangular form and then bonding the layers together. It is then cut in two, forming two C shapes, and the core assembled by binding the two C halves together with a steel strap. They have the advantage that the flux is always oriented parallel to the metal grains, reducing reluctance.

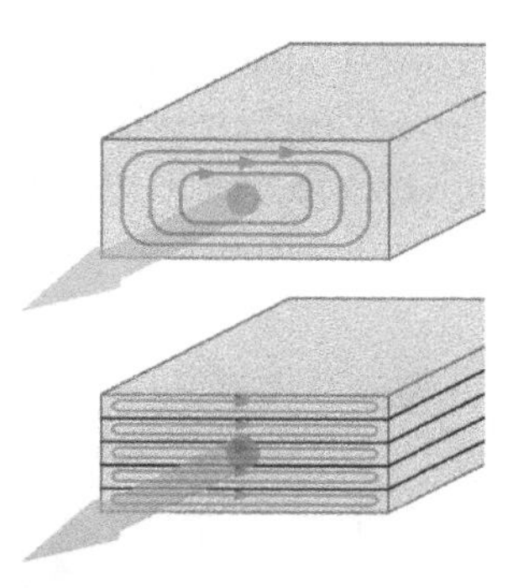

Laminating the core greatly reduces eddy-current losses

A steel core's remanence means that it retains a static magnetic field when power is removed. When power is then reapplied, the residual field will cause a high inrush current until the effect of the remaining magnetism is reduced, usually after a few cycles of the applied AC waveform. Overcurrent protection devices such as fuses must be selected to allow this harmless inrush to pass. On transformers connected to long, overhead power transmission lines, induced currents due to geomagnetic disturbances during solar storms can cause saturation of the core and operation of transformer protection devices.

Distribution transformers can achieve low no-load losses by using cores made with low-loss high-permeability silicon steel or amorphous (non-crystalline) metal alloy. The higher initial cost of the core material is offset over the life of the transformer by its lower losses at light load.

Solid Cores

Powdered iron cores are used in circuits such as switch-mode power supplies that operate above mains frequencies and up to a few tens of kilohertz. These materials combine high magnetic permeability with high bulk electrical resistivity. For frequencies extending beyond the VHF band, cores made from non-conductive magnetic ceramic materials called ferrites are common. Some radio-frequency transformers also have movable cores (sometimes called 'slugs') which allow adjustment of the coupling coefficient (and bandwidth) of tuned radio-frequency circuits.

Toroidal Cores

Toroidal transformers are built around a ring-shaped core, which, depending on operating frequency, is made from a long strip of silicon steel or permalloy wound into a coil, powdered iron, or ferrite. A strip construction ensures that the grain boundaries are optimally aligned, improving the transformer's efficiency by reducing the core's reluctance. The closed ring shape eliminates air gaps inherent in the construction of an E-I core. The cross-section of the ring is usually square or rectangular, but more expensive cores with circular cross-sections are also available. The primary and secondary coils are often wound concentrically to cover the entire surface of the core. This minimizes the length of wire needed and provides screening to minimize the core's magnetic field from generating electromagnetic interference.

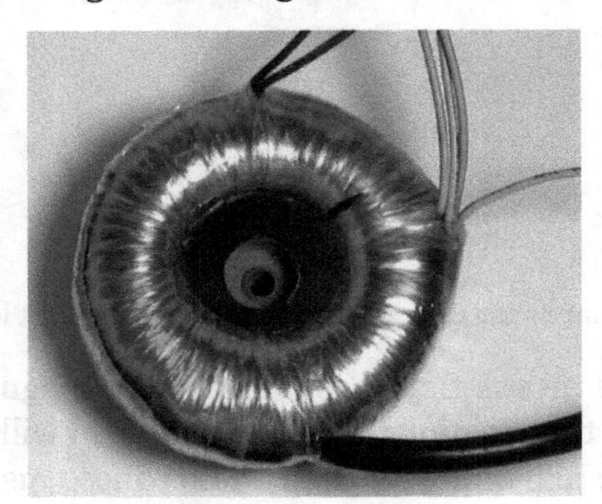

Small toroidal core transformer

Toroidal transformers are more efficient than the cheaper laminated E-I types for a similar power level. Other advantages compared to E-I types, include smaller size (about half), lower weight (about half), less mechanical hum (making them superior in audio amplifiers), lower exterior magnetic field (about one tenth), low off-load losses (making them more efficient in standby circuits), single-bolt mounting, and greater choice of shapes. The main disadvantages are higher cost and limited power capacity. Because of the lack of a residual gap in the magnetic path, toroidal transformers also tend to exhibit higher inrush current, compared to laminated E-I types.

Ferrite toroidal cores are used at higher frequencies, typically between a few tens of kilohertz to hundreds of megahertz, to reduce losses, physical size, and weight of inductive components. A drawback of toroidal transformer construction is the higher labor cost of winding. This is because it is necessary to pass the entire length of a coil winding through the core aperture each time a single turn is added to the coil. As a consequence, toroidal transformers rated more than a few kVA are uncommon. Relatively few toroids are offered with power ratings above 10 kVA, and practically none above 25 kVA. Small distribution transformers may achieve some of the benefits of a toroidal core by splitting it and forcing it open, then inserting a bobbin containing primary and secondary windings.

Air Cores

A physical core is not an absolute requisite and a functioning transformer can be produced simply by placing the windings near each other, an arrangement termed an 'air-core' transformer. The air which comprises the magnetic circuit is essentially lossless, and so an air-core transformer eliminates loss due to hysteresis in the core material. The leakage inductance is inevitably high, resulting in very poor regulation, and so such designs are unsuitable for use in power distribution. They have however very high bandwidth, and are frequently employed in radio-frequency applications, for which a satisfactory coupling coefficient is maintained by carefully overlapping the primary and secondary windings. They're also used for resonant transformers such as Tesla coils where they can achieve reasonably low loss in spite of the high leakage inductance.

Windings

The conducting material used for the windings depends upon the application, but in all cases the individual turns must be electrically insulated from each other to ensure that the current travels throughout every turn. For small power and signal transformers, in which currents are low and the potential difference between adjacent turns is small, the coils are often wound from enamelled magnet wire, such as Formvar wire. Larger power transformers operating at high voltages may be wound with copper rectangular strip conductors insulated by oil-impregnated paper and blocks of pressboard.

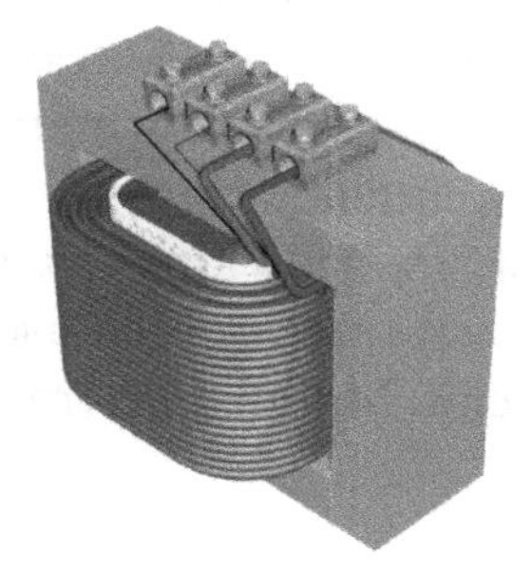

Windings are usually arranged concentrically to minimize flux leakage.

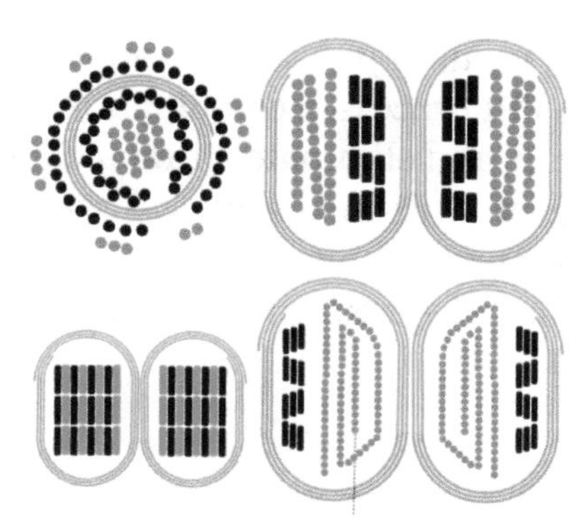

Cut view through transformer windings. White: insulator. Green spiral: Grain oriented silicon steel. Black: Primary winding made of oxygen-free copper. Red: Secondary winding. Top left: Toroidal transformer. Right: C-core, but E-core would be similar. The black windings are made of film. Top: Equally low capacitance between all ends of both windings. Since most cores are at least moderately conductive they also need insulation. Bottom: Lowest capacitance for one end of the secondary winding needed for low-power high-voltage transformers. Bottom left: Reduction of leakage inductance would lead to increase of capacitance.

High-frequency transformers operating in the tens to hundreds of kilohertz often have windings made of braided Litz wire to minimize the skin-effect and proximity effect losses. Large power transformers use multiple-stranded conductors as well, since even at low power frequencies non-uniform distribution of current would otherwise exist in high-current windings. Each strand is individually insulated, and the strands are arranged so that at certain points in the winding, or throughout the whole winding, each portion occupies different relative positions in the complete conductor. The transposition equalizes the current flowing in each strand of the conductor, and reduces eddy current losses in the winding itself. The stranded conductor is also more flexible than a solid conductor of similar size, aiding manufacture.

The windings of signal transformers minimize leakage inductance and stray capacitance to improve high-frequency response. Coils are split into sections, and those sections interleaved between the sections of the other winding.

Power-frequency transformers may have taps at intermediate points on the winding, usually on the higher voltage winding side, for voltage adjustment. Taps may be manually reconnected, or a manual or automatic switch may be provided for changing taps. Automatic on-load tap changers are used in electric power transmission or distribution, on equipment such as arc furnace transformers, or for automatic voltage regulators for sensitive loads. Audio-frequency transformers, used for the distribution of audio to public address loudspeakers, have taps to allow adjustment of impedance to each speaker. A center-tapped transformer is often used in the output stage of an audio power amplifier in a push-pull circuit. Modulation transformers in AM transmitters are very similar.

Dry-type transformer winding insulation systems can be either of standard open-wound 'dip-and-bake' construction or of higher quality designs that include vacuum pressure impregnation (VPI), vacuum pressure encapsulation (VPE), and cast coil encapsulation processes. In the VPI process, a combination of heat, vacuum and pressure is used to thoroughly seal, bind, and eliminate entrained air voids in the winding polyester resin insulation coat layer, thus increasing resistance to corona. VPE windings are similar to VPI windings but provide more protection against environmental effects, such as from water, dirt or corrosive ambients, by multiple dips including typically in terms of final epoxy coat.

Cooling

To place the cooling problem in perspective, the accepted rule of thumb is that the life expectancy of insulation in all electrics, including all transformers, is halved for about every 7 °C to 10 °C increase in operating temperature, this life expectancy halving rule holding more narrowly when the increase is between about 7 °C to 8 °C in the case of transformer winding cellulose insulation.

Small dry-type and liquid-immersed transformers are often self-cooled by natural convection and radiation heat dissipation. As power ratings increase, transformers are often cooled by forced-air cooling, forced-oil cooling, water-cooling, or combinations of these. Large transformers are filled with transformer oil that both cools and insulates the windings. Transformer oil is a highly refined mineral oil that cools the windings and insulation by circulating within the transformer tank. The mineral oil and paper insulation system has been extensively studied and used for more than 100 years. It is estimated that 50% of power transformers will survive 50 years of use, that the average age of failure of power transformers is about 10 to 15 years, and that about 30% of power transformer failures are due to insulation and overloading failures. Prolonged operation at elevated temperature degrades insulating properties of winding insulation and dielectric coolant, which not only shortens transformer life but can ultimately lead to catastrophic transformer failure. With a great body of empirical study as a guide, transformer oil testing including dissolved gas analysis provides valuable maintenance information. This underlines the need to monitor, model, forecast and manage oil and winding conductor insulation temperature conditions under varying, possibly difficult, power loading conditions.

Cutaway view of liquid-immersed construction transformer. The conservator (reservoir) at top provides liquid-to-atmosphere isolation as coolant level and temperature changes. The walls and fins provide required heat dissipation balance.

Building regulations in many jurisdictions require indoor liquid-filled transformers to either use dielectric fluids that are less flammable than oil, or be installed in fire-resistant rooms. Air-cooled dry transformers can be more economical where they eliminate the cost of a fire-resistant transformer room.

The tank of liquid filled transformers often has radiators through which the liquid coolant circulates by natural convection or fins. Some large transformers employ electric fans for forced-air cooling, pumps for forced-liquid cooling, or have heat exchangers for water-cooling. An oil-immersed transformer may be equipped with a Buchholz relay, which, depending on severity of gas accumulation due to internal arcing, is used to either alarm or de-energize the transformer. Oil-immersed transformer installations usually include fire protection measures such as walls, oil containment, and fire-suppression sprinkler systems.

Polychlorinated biphenyls have properties that once favored their use as a dielectric coolant, though concerns over their environmental persistence led to a widespread ban on their use. Today, non-toxic, stable silicone-based oils, or fluorinated hydrocarbons may be used where the expense of a fire-resistant liquid offsets additional building cost for a transformer vault. PCBs for new equipment were banned in 1981 and in 2000 for use in existing equipment in United Kingdom Legislation enacted in Canada between 1977 and 1985 essentially bans PCB use in transformers manufactured in or imported into the country after 1980, the maximum allowable level of PCB contamination in existing mineral oil transformers being 50 ppm.

Some transformers, instead of being liquid-filled, have their windings enclosed in sealed, pressurized tanks and cooled by nitrogen or sulfur hexafluoride gas.

Experimental power transformers in the 500-to-1,000 kVA range have been built with liquid nitrogen or helium cooled superconducting windings, which eliminates winding losses without affecting core losses.

Insulation Drying

Construction of oil-filled transformers requires that the insulation covering the windings be thoroughly dried of residual moisture before the oil is introduced. Drying is carried out at the factory, and may also be required as a field service. Drying may be done by circulating hot air around the core, or by vapor-phase drying (VPD) where an evaporated solvent transfers heat by condensation on the coil and core.

For small transformers, resistance heating by injection of current into the windings is used. The heating can be controlled very well, and it is energy efficient. The method is called low-frequency heating (LFH) since the current used is at a much lower frequency than that of the power grid, which is normally 50 or 60 Hz. A lower frequency reduces the effect of inductance, so the voltage required can be reduced. The LFH drying method is also used for service of older transformers.

Bushings

Larger transformers are provided with high-voltage insulated bushings made of polymers or porcelain. A large bushing can be a complex structure since it must provide careful control of the electric field gradient without letting the transformer leak oil.

Classification Parameters

Transformers can be classified in many ways, such as the following:

- Power capacity: From a fraction of a volt-ampere (VA) to over a thousand MVA.
- Duty of a transformer: Continuous, short-time, intermittent, periodic, varying.
- Frequency range: Power-frequency, audio-frequency, or radio-frequency.
- Voltage class: From a few volts to hundreds of kilovolts.
- Cooling type: Dry and liquid-immersed – self-cooled, forced air-cooled; liquid-immersed – forced oil-cooled, water-cooled.
- Circuit application: Such as power supply, impedance matching, output voltage and current stabilizer or circuit isolation.
- Utilization: Pulse, power, distribution, rectifier, arc furnace, amplifier output, etc..
- Basic magnetic form: Core form, shell form.
- Constant-potential transformer descriptor: Step-up, step-down, isolation.
- General winding configuration: By EIC vector group – various possible two-winding combinations of the phase designations delta, wye or star, and zigzag or interconnected star; other – autotransformer, Scott-T, zigzag grounding transformer winding.
- Rectifier phase-shift winding configuration: 2-winding, 6-pulse; 3-winding, 12-pulse; . . . n-winding, [n-1]*6-pulse; polygon; etc..

Types

Various specific electrical application designs require a variety of transformer types. Although they all share the basic characteristic transformer principles, they are customize in construction or electrical properties for certain installation requirements or circuit conditions.

- Autotransformer: Transformer in which part of the winding is common to both primary and secondary circuits, leading to increased efficiency, smaller size, and a higher degree of voltage regulation.
- Capacitor voltage transformer: Transformer in which capacitor divider is used to reduce high voltage before application to the primary winding.
- Distribution transformer, power transformer: International standards make a distinction in terms of distribution transformers being used to distribute energy from transmission lines and networks for local consumption and power transformers being used to transfer electric energy between the generator and distribution primary circuits.

- Phase angle regulating transformer: A specialised transformer used to control the flow of real power on three-phase electricity transmission networks.
- Scott-T transformer: Transformer used for phase transformation from three-phase to two-phase and vice versa.
- Polyphase transformer: Any transformer with more than one phase.
- Grounding transformer: Transformer used for grounding three-phase circuits to create a neutral in a three wire system, using a wye-delta transformer, or more commonly, a zigzag grounding winding.
- Leakage transformer: Transformer that has loosely coupled windings.
- Resonant transformer: Transformer that uses resonance to generate a high secondary voltage.
- Audio transformer: Transformer used in audio equipment.
- Output transformer: Transformer used to match the output of a valve amplifier to its load.
- Instrument transformer: Potential or current transformer used to accurately and safely represent voltage, current or phase position of high voltage or high power circuits.
- Pulse transformer: Specialized small-signal transformer used to transmit digital signaling while providing electrical isolation, commonly used in Ethernet computer networks as 10BASE-T, 100BASE-T and 1000BASE-T.

An electrical substation in Melbourne, Australia showing three of five 220 kV – 66 kV transformers, each with a capacity of 150 MVA

Applications

Transformers are used to increase (or step-up) voltage before transmitting electrical energy over long distances through wires. Wires have resistance which loses energy through joule heating at a rate corresponding to square of the current. By transforming power to a higher voltage transformers enable economical transmission of power and distribution. Consequently, transformers have shaped the electricity supply industry, permitting generation to be located remotely from points of demand. All but a tiny fraction of the world's electrical power has passed through a series of transformers by the time it reaches the consumer.

Transformer at the Limestone Generating Station in Manitoba, Canada

Since the high voltages carried in the wires are significantly greater than what is needed in-home, transformers are also used extensively in electronic products to decrease (or step-down) the supply voltage to a level suitable for the low voltage circuits they contain. The transformer also electrically isolates the end user from contact with the supply voltage.

Signal and audio transformers are used to couple stages of amplifiers and to match devices such as microphones and record players to the input of amplifiers. Audio transformers allowed telephone circuits to carry on a two-way conversation over a single pair of wires. A balun transformer converts a signal that is referenced to ground to a signal that has balanced voltages to ground, such as between external cables and internal circuits. Transformers made to medical grade standards isolate the users from the direct current. These are found commonly used in conjunction with hospital beds, dentist chairs, and other medical lab equipment.

History

Discovery of Induction

Electromagnetic induction, the principle of the operation of the transformer, was discovered independently by Michael Faraday in 1831, Joseph Henry in 1832, and others. The relationship between EMF and magnetic flux is an equation now known as Faraday's law of induction:

$$|\mathcal{E}| = \left|\frac{d\Phi_B}{dt}\right|.$$

where $|\mathcal{E}|$ is the magnitude of the EMF in Volts and Φ_B is the magnetic flux through the circuit in webers.

Preceded by Francesco Zantedeschi in 1830, Faraday performed early experiments on induction between coils of wire, including winding a pair of coils around an iron ring, thus creating the first toroidal closed-core transformer. However he only applied individual pulses of current to his transformer, and never discovered the relation between the turns ratio and EMF in the windings. Presently, toroidal power transformers are manufactured with dual primary coils and dual secondary coils, allowing for a high degree of input and output voltage flexibility.

The first type of transformer to see wide use was the induction coil, invented by Rev. Nicholas Callan of Maynooth College, Ireland in 1836. He was one of the first researchers to realize the more turns the secondary winding has in relation to the primary winding, the larger the induced secondary EMF will be. Induction coils evolved from scientists' and inventors' efforts to get higher voltages from batteries. Since batteries produce direct current (DC) rather than AC, induction coils relied upon vibrating electrical contacts that regularly interrupted the current in the primary to create the flux changes necessary for induction. Between the 1830s and the 1870s, efforts to build better induction coils, mostly by trial and error, slowly revealed the basic principles of transformers.

Induction coil, 1900, Bremerhaven, Germany

Induction Coils

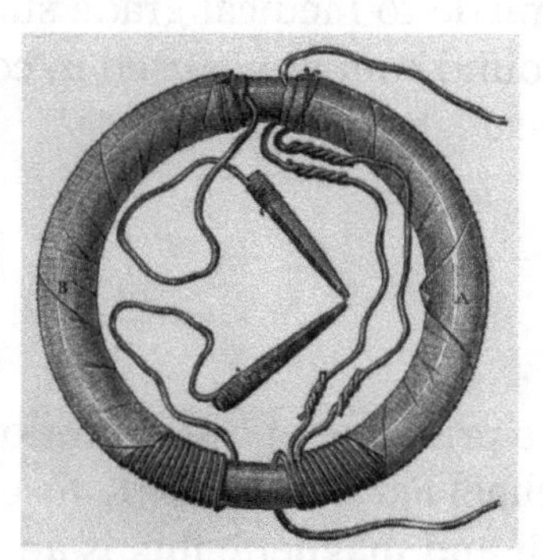

Faraday's ring transformer

First Alternating Current Transformers

By the 1870s, efficient generators producing alternating current (AC) were available, and it was found AC could power an induction coil directly, without an interrupter.

In 1876, Russian engineer Pavel Yablochkov invented a lighting system based on a set of induction coils where the primary windings were connected to a source of AC. The secondary windings could be connected to several 'electric candles' (arc lamps) of his own design. The coils Yablochkov employed functioned essentially as transformers.

In 1878, the Ganz factory, Budapest, Hungary, began equipment for electric lighting and, by 1883, had installed over fifty systems in Austria-Hungary. Their AC systems used arc and incandescent lamps, generators, and other equipment.

Lucien Gaulard and John Dixon Gibbs first exhibited a device with an open iron core called a 'secondary generator' in London in 1882, then sold the idea to the Westinghouse company in the United States. They also exhibited the invention in Turin, Italy in 1884, where it was adopted for an electric lighting system. However, the efficiency of their open-core bipolar apparatus remained very low.

Early Series Circuit Transformer Distribution

Induction coils with open magnetic circuits are inefficient at transferring power to loads. Until about 1880, the paradigm for AC power transmission from a high voltage supply to a low voltage load was a series circuit. Open-core transformers with a ratio near 1:1 were connected with their primaries in series to allow use of a high voltage for transmission while presenting a low voltage to the lamps. The inherent flaw in this method was that turning off a single lamp (or other electric device) affected the voltage supplied to all others on the same circuit. Many adjustable transformer designs were introduced to compensate for this problematic characteristic of the series circuit, including those employing methods of adjusting the core or bypassing the magnetic flux around part of a coil. Efficient, practical transformer designs did not appear until the 1880s, but within a decade, the transformer would be instrumental in the War of Currents, and in seeing AC distribution systems triumph over their DC counterparts, a position in which they have remained dominant ever since.

Shell form transformer. Sketch used by Uppenborn to describe ZBD engineers' 1885 patents and earliest articles.

Core form, front; shell form, back. Earliest specimens of ZBD-designed high-efficiency constant-potential transformers manufactured at the Ganz factory in 1885.

Closed-core Transformers and Parallel Power Distribution

In the autumn of 1884, Károly Zipernowsky, Ottó Bláthy and Miksa Déri (ZBD), three engineers associated with the Ganz factory, had determined that open-core devices were impracticable, as they were incapable of reliably regulating voltage. In their joint 1885 patent applications for novel transformers (later called ZBD transformers), they described two designs with closed magnetic circuits where copper windings were either a) wound around iron wire ring core or b) surrounded by iron wire core. The two designs were the first application of the two basic transformer constructions in common use to this day, which can as a class all be termed as either core form or shell form (or alternatively, core type or shell type), as in a) or b), respectively. The Ganz factory had also in the autumn of 1884 made delivery of the world's first five high-efficiency AC transformers, the first of these units having been shipped on September 16, 1884. This first unit had been manufactured to the following specifications: 1,400 W, 40 Hz, 120:72 V, 11.6:19.4 A, ratio 1.67:1, one-phase, shell form.

In both designs, the magnetic flux linking the primary and secondary windings traveled almost entirely within the confines of the iron core, with no intentional path through air. The new transformers were 3.4 times more efficient than the open-core bipolar devices of Gaulard and Gibbs. The ZBD patents included two other major interrelated innovations: one concerning the use of parallel connected, instead of series connected, utilization loads, the other concerning the ability to have high turns ratio transformers such that the supply network voltage could be much higher (initially 1,400 to 2,000 V) than the voltage of utilization loads (100 V initially preferred). When employed in parallel connected electric distribution systems, closed-core transformers finally made it technically and economically feasible to provide electric power for lighting in homes, businesses and public spaces. Bláthy had suggested the use of closed cores, Zipernowsky had suggested the use of parallel shunt connections, and Déri had performed the experiments;

Transformers today are designed on the principles discovered by the three engineers. They also popularized the word 'transformer' to describe a device for altering the EMF of an electric current, although the term had already been in use by 1882. In 1886, the ZBD engineers designed, and the Ganz factory supplied electrical equipment for, the world's first power station that used AC generators to power a parallel connected common electrical network, the steam-powered Rome-Cerchi power plant.

Although George Westinghouse had bought Gaulard and Gibbs' patents in 1885, the Edison Electric Light Company held an option on the US rights for the ZBD transformers, requiring Westinghouse to pursue alternative designs on the same principles. He assigned to William Stanley the task of developing a device for commercial use in United States. Stanley's first patented design was for induction coils with single cores of soft iron and adjustable gaps to regulate the EMF present in the secondary winding. This design was first used commercially in the US in 1886 but Westinghouse was intent on improving the Stanley design to make it (unlike the ZBD type) easy and cheap to produce.

Westinghouse, Stanley and associates soon developed an easier to manufacture core, consisting of a stack of thin 'Eshaped' iron plates, insulated by thin sheets of paper or other insulating material. Prewound copper coils could then be slid into place, and straight iron plates laid in to create a closed magnetic circuit. Westinghouse applied for a patent for the new low-cost design in December 1886; it was granted in July 1887.

Other Early Transformer Designs

In 1889, Russian-born engineer Mikhail Dolivo-Dobrovolsky developed the first three-phase transformer at the Allgemeine Elektricitäts-Gesellschaft ('General Electricity Company') in Germany.

In 1891, Nikola Tesla invented the Tesla coil, an air-cored, dual-tuned resonant transformer for producing very high voltages at high frequency.

Fault Current Limiter

A fault current limiter (FCL), or fault current controller (FCC), is a device which limits the prospective fault current when a fault occurs (e.g. in a power transmission network) without complete disconnection. The term includes superconducting, solid-state and inductive devices.

Applications

Electric power distribution systems include circuit breakers to disconnect power in case of a fault, but to maximize reliability, they wish to disconnect the smallest possible portion of the network. This means that even the smallest circuit breakers, as well as all wiring to them, must be able to disconnect large fault currents.

A problem arises if the electricity supply is upgraded, by adding new generation capacity or by adding cross-connections. Because these increase the amount of power that can be supplied, all of the branch circuits must have their bus bars and circuit breakers upgraded to handle the new higher fault current limit.

This poses a particular problem when distributed generation, such as wind farms and rooftop solar power, is added to an existing electric grid. It is desirable to be able to add additional power sources without large system-wide upgrades.

A simple solution is to add electrical impedance to the circuit. This limits the rate at which current can increase, which limits the level the fault current can rise to before the breaker is opened. However, this also limits the ability of the circuit to satisfy rapidly changing demand, so the addition or removal of large loads causes unstable power.

A fault current limiter is a nonlinear element which has a low impedance at normal current levels, but presents a higher impedance at fault current levels. Further, this change is extremely rapid, before a circuit breaker can trip a few millisecond later. (High-power circuit breakers are synchronized to the alternating current zero crossing to minimize arcing.)

While the power is unstable during the fault, it is not completely disconnected. After the faulting branch is disconnected, the fault current limiter automatically returns to normal operation.

Superconducting Fault Current Limiter

Superconducting fault current limiters exploit the extremely rapid loss of superconductivity (called "quenching) above a critical combination of temperature, current density, and magnetic field. In normal operation, current flows through the superconductor without resistance and negligible impedance.

If a fault develops, the superconductor quenches, its resistance rises sharply, and current is diverted to a parallel circuit with the desired higher impedance.

(The structure is not usable as a circuit breaker, because the normally-conducting superconductive material does not have a high enough resistance. It is only high enough to cause sufficient heating to melt the material.)

Superconducting fault current limiters are described as being in one of two major categories: resistive or inductive.

In a resistive FCL, the current passes directly through the superconductor. When it quenches, the sharp rise in resistance reduces the fault current from what it would otherwise be (the prospective fault current). A resistive FCL can be either DC or AC. If it is AC, then there will be a steady power dissipation from AC losses (superconducting hysteresis losses) which must be removed by the cryogenic system. An AC FCL is usually made from wire wound non-inductively; otherwise the inductance of the device would create an extra constant power loss on the system.

Inductive FCLs come in many variants, but the basic concept is a transformer with a resistive FCL as the secondary. In un-faulted operation, there is no resistance in the secondary and so the inductance of the device is low. A fault current quenches the superconductor, the secondary becomes resistive and the inductance of the whole device rises. The advantage of this design is that there is no heat ingress through current leads into the superconductor, and so the cryogenic power load may be lower. However, the large amount of iron required means that inductive FCLs are much bigger and heavier than resistive FCLs.

The quench process is a two-step process. First, a small region quenches directly in response to a high current density. This section rapidly heats by Joule heating, and the increase in temperature quenches adjacent regions.GridON Ltd has developed the first commercial inductive FCL for distribution & transmission networks. Using a unique and proprietary concept of magnetic-flux alteration - requiring no superconducting or cryogenic components - the self-triggered FCL instantaneously increases its impedance tenfold upon fault condition. It limits the fault current for its entire duration and recovers to its normal condition immediately thereafter. This inductive FCL is scalable to extra high voltage ratings.

Development of the Superconducting Fault Current Limiters

FCLs are under active development. In 2007, there were at least six national and international projects using magnesium diboride wire or YBCO tape, and two using BSCCO-2212 rods. Countries active in FCL development are Germany, the UK, the USA, Korea and China. In 2007, the US Department of Energy spent $29m on three FCL development projects.

High temperature superconductors are required for practical FCLs. AC losses generate constant heat inside the superconductor, and the cost of cryogenic cooling at liquid helium temperatures required by low temperature superconductors makes the whole device uneconomic.

First applications for FCLs are likely to be used to help control medium-voltage electricity distribution systems, followed by electric-drive ships: naval vessels, submarines and cruise ships. Larger FCLs may eventually be deployed in high-voltage transmission systems.

References

- Ned Mohan; T. M. Undeland; William P. Robbins (2003). Power Electronics: Converters, Applications, and Design. United States of America: John Wiley & Sons, Inc. ISBN 0-471-22693-9.
- Chapman, Stephen (2002). Electric Machinery and Power System Fundamentals. Boston: McGraw-Hill. pp. Chapters 6 and 7. ISBN 0-07-229135-4.
- Grigsby, Leonard (2007). Electric Power Generation, Transmission, and Distribution. CRC Press 2007. pp. Chapter 14. ISBN 978-0-8493-9292-4.
- Short, T.A. (2014). Electric Power Distribution Handbook. Boca Raton, Florida, USA: CRC Press. pp. 1–33. ISBN 978-1-4665-9865-2.
- Abdelhay A. Sallam and Om P. Malik (May 2011). Electric Distribution Systems. IEEE Computer Society Press. p. 21. ISBN 9780470276822.
- Hans Dieter Betz, Ulrich Schumann, Pierre Laroche (2009). Lightning: Principles, Instruments and Applications. Springer, pp. 202–203. ISBN 978-1-4020-9078-3. Retrieved on May 13, 2009.
- Thomas P. Hughes (1993). Networks of Power: Electrification in Western Society, 1880–1930. Baltimore: Johns Hopkins University Press. pp. 119–122. ISBN 0-8018-4614-5.
- Mack, James E.; Shoemaker, Thomas (2006). Chapter 15 - Distribution Transformers (PDF) (11th ed.). New York: McGraw-Hill. pp. 15–1 to 15–22. ISBN 0-07-146789-0.
- Del Vecchio, Robert M.; et al. (2002). Transformer Design Principles: With Applications to Core-Form Power Transformers. Boca Raton: CRC Press. pp. 10–11, Fig. 1.8. ISBN 90-5699-703-3.
- Fink, Donald G.; Beatty, H. Wayne (Eds.) (1978). Standard Handbook for Electrical Engineers (11th ed.). McGraw Hill. pp. 10–38 through 10–40. ISBN 978-0-07-020974-9.
- Ryan, Hugh M. (2001). High Voltage Engineering and Testing. Institution Electrical Engineers. pp. 416–417. ISBN 0-85296-775-6.
- Chow, Tai L. (2006). Introduction to Electromagnetic Theory: A Modern Perspective. Sudbury, Mass.: Jones and Bartlett Publishers. p. 171. ISBN 0-7637-3827-1.
- Hughes, Thomas P. (1993). Networks of Power: Electrification in Western Society, 1880-1930. Baltimore: The Johns Hopkins University Press. p. 96. ISBN 0-8018-2873-2. Retrieved Sep 9, 2009.
- Smil, Vaclav (2005). Creating the Twentieth Century: Technical Innovations of 1867—1914 and Their Lasting Impact. Oxford: Oxford University Press. p. 71. ISBN 978-0-19-803774-3.
- Skrabec, Quentin R. (2007). George Westinghouse: Gentle Genius. Algora Publishing. p. 102. ISBN 978-0-87586-508-9.
- Neidhöfer, Gerhard (2008). Michael von Dolivo-Dobrowolsky and Three-Phase: The Beginnings of Modern e Technology and Power Supply (in German). In collaboration with VDE "History of Electrical Engineering" Committee (2 ed.). Berlin: VDE-Verl. ISBN 978-3-8007-3115-2.
- Chapman, Stephen (2002). Electric Machinery and Power System Fundamentals. Boston: McGraw-Hill. pp. Chapter 4. ISBN 0-07-229135-4.
- Kulkarni, S. V.; Khaparde, S. A. (May 24, 2004). Transformer Engineering: Design and Practice. CRC. pp. 36–37. ISBN 0-8247-5653-3.
- "Toroidal Line Power Transformers. Power Ratings Tripled. | Magnetics Magazine". www.magneticsmagazine.com. Retrieved 2016-09-23.

5

Analysis of Power Engineering

Power engineering uses numerical analysis in the form of power-flow study to assess various parameters of AC current like voltage, real power, voltage angles and reactive power. It uses notations like one-line diagram to achieve this. The content of the chapter is designed to further the reader's comprehension of power systems. Power engineering is best understood in confluence with the major topics listed in the following chapter.

Power-flow Study

In power engineering, the power-flow study, or load-flow study, is a numerical analysis of the flow of electric power in an interconnected system. A power-flow study usually uses simplified notation such as a one-line diagram and per-unit system, and focuses on various aspects of AC power parameters, such as voltages, voltage angles, real power and reactive power. It analyzes the power systems in normal steady-state operation.

Power-flow or load-flow studies are important for planning future expansion of power systems as well as in determining the best operation of existing systems. The principal information obtained from the power-flow study is the magnitude and phase angle of the voltage at each bus, and the real and reactive power flowing in each line.

Commercial power systems are usually too complex to allow for hand solution of the power flow. Special purpose network analyzers were built between 1929 and the early 1960s to provide laboratory-scale physical models of power systems. Large-scale digital computers replaced the analog methods with numerical solutions.

In addition to a power-flow study, computer programs perform related calculations such as short-circuit fault analysis, stability studies (transient & steady-state), unit commitment and economic dispatch. In particular, some programs use linear programming to find the optimal power flow, the conditions which give the lowest cost per kilowatt hour delivered.

A load flow study is especially valuable for a system with multiple load centers, such as a refinery complex. The power flow study is an analysis of the system's capability to adequately supply the connected load. The total system losses, as well as individual line losses, also are tabulated. Transformer tap positions are selected to ensure the correct voltage at critical locations such as motor control centers. Performing a load flow study on an existing system provides insight and recommendations as to the system operation and optimization of control settings to obtain maximum capacity while minimizing the operating costs. The results of such an analysis are in terms of active power, reactive power, magnitude and phase angle. Furthermore, power-flow computations are crucial for optimal operations of groups of generating units.

The Open Energy Modelling Initiative promotes open source load-flow models, and other types of energy system models.

Model

An alternating current power-flow model is a model used in electrical engineering to analyze power grids. It provides a nonlinear system which describes the energy flow through each transmission line. The problem is non-linear because the power flow into load impedances is a function of the square of the applied voltages. Due to nonlinearity, in many cases the analysis of large network via AC power-flow model is not feasible, and a linear (but less accurate) DC power-flow model is used instead.

Usually analysis of a three-phase system is simplified by assuming balanced loading of all three phases. Steady-state operation is assumed, with no transient changes in power flow or voltage due to load or generation changes. The system frequency is also assumed to be constant. A further simplification is to use the per-unit system to represent all voltages, power flows, and impedances, scaling the actual target system values to some convenient base. A system one-line diagram is the basis to build a mathematical model of the generators, loads, buses, and transmission lines of the system, and their electrical impedances and ratings.

Power-flow Problem Formulation

The goal of a power-flow study is to obtain complete voltages angle and magnitude information for each bus in a power system for specified load and generator real power and voltage conditions. Once this information is known, real and reactive power flow on each branch as well as generator reactive power output can be analytically determined. Due to the nonlinear nature of this problem, numerical methods are employed to obtain a solution that is within an acceptable tolerance.

The solution to the power-flow problem begins with identifying the known and unknown variables in the system. The known and unknown variables are dependent on the type of bus. A bus without any generators connected to it is called a Load Bus. With one exception, a bus with at least one generator connected to it is called a Generator Bus. The exception is one arbitrarily-selected bus that has a generator. This bus is referred to as the slack bus.

In the power-flow problem, it is assumed that the real power P_D and reactive power Q_D at each Load Bus are known. For this reason, Load Buses are also known as PQ Buses. For Generator Buses, it is assumed that the real power generated P_G and the voltage magnitude |V| is known. For the Slack Bus, it is assumed that the voltage magnitude |V| and voltage phase Θ are known. Therefore, for each Load Bus, both the voltage magnitude and angle are unknown and must be solved for; for each Generator Bus, the voltage angle must be solved for; there are no variables that must be solved for the Slack Bus. In a system with N buses and R generators, there are then $2(N-1)-(R-1)$ unknowns.

In order to solve for the $2(N-1)-(R-1)$ unknowns, there must be $2(N-1)-(R-1)$ equations that do not introduce any new unknown variables. The possible equations to use are power balance equations, which can be written for real and reactive power for each bus. The real power balance equation is:

$$0=-P_i+\sum_{k=1}^{N}|V_i||V_k|(G_{ik}\cos\theta_{ik}+B_{ik}\sin\theta_{ik})$$

where P_i is the net power injected at bus i, G_{ik} is the real part of the element in the bus admittance matrix Y_{BUS} corresponding to the ith row and kth column, B_{ik} is the imaginary part of the element in the Y_{BUS} corresponding to the ith row and kth column and θ_{ik} is the difference in voltage angle between the ith and kth buses ($\theta_{ik}=\delta_i-\delta_k$). The reactive power balance equation is:

$$0=-Q_i+\sum_{k=1}^{N}|V_i||V_k|(G_{ik}\sin\theta_{ik}-B_{ik}\cos\theta_{ik})$$

where Q_i is the net reactive power injected at bus i.

Equations included are the real and reactive power balance equations for each Load Bus and the real power balance equation for each Generator Bus. Only the real power balance equation is written for a Generator Bus because the net reactive power injected is assumed to be unknown and therefore including the reactive power balance equation would result in an additional unknown variable. For similar reasons, there are no equations written for the Slack Bus.

In many transmission systems, the voltage angles θ_{ik} are usually relatively small. There is thus a strong coupling between real power and voltage angle, and between reactive power and voltage magnitude, while the coupling between real power and voltage magnitude, as well as reactive power and voltage angle, is weak. As a result, real power is usually transmitted from the bus with higher voltage angle to the bus with lower voltage angle, and reactive power is usually transmitted from the bus with higher voltage magnitude to the bus with lower voltage magnitude. However, this approximation does not hold when the voltage angle is very large.

Newton–raphson Solution Method

There are several different methods of solving the resulting nonlinear system of equations. The most popular is known as the Newton–Raphson method. This method begins with initial guesses of all unknown variables (voltage magnitude and angles at Load Buses and voltage angles at Generator Buses). Next, a Taylor Series is written, with the higher order terms ignored, for each of the power balance equations included in the system of equations . The result is a linear system of equations that can be expressed as:

$$\begin{bmatrix} \Delta\theta \\ \Delta|V| \end{bmatrix} = -J^{-1} \begin{bmatrix} \Delta P \\ \Delta Q \end{bmatrix}$$

where ΔP and ΔQ are called the mismatch equations:

$$\Delta P_i = -P_i + \sum_{k=1}^{N} |V_i||V_k|(G_{ik}\cos\theta_{ik} + B_{ik}\sin\theta_{ik})$$

$$\Delta Q_i = -Q_i + \sum_{k=1}^{N} |V_i||V_k|(G_{ik}\sin\theta_{ik} - B_{ik}\cos\theta_{ik})$$

and J is a matrix of partial derivatives known as a Jacobian: $J = \begin{bmatrix} \frac{\partial\Delta P}{\partial\theta} & \frac{\partial\Delta P}{\partial|V|} \\ \frac{\partial\Delta Q}{\partial\theta} & \frac{\partial\Delta Q}{\partial|V|} \end{bmatrix}$..

The linearized system of equations is solved to determine the next guess (m + 1) of voltage magnitude and angles based on:

$$\theta^{m+1} = \theta^m + \Delta\theta$$

$$|V|^{m+1} = |V|^m + \Delta|V|$$

The process continues until a stopping condition is met. A common stopping condition is to terminate if the norm of the mismatch equations is below a specified tolerance.

A rough outline of solution of the power-flow problem is:

1. Make an initial guess of all unknown voltage magnitudes and angles. It is common to use a “flat start” in which all voltage angles are set to zero and all voltage magnitudes are set to 1.0 p.u.
2. Solve the power balance equations using the most recent voltage angle and magnitude values.
3. Linearize the system around the most recent voltage angle and magnitude values
4. Solve for the change in voltage angle and magnitude
5. Update the voltage magnitude and angles
6. Check the stopping conditions, if met then terminate, else go to step 2.

Other Power-flow Methods

- Gauss–Seidel method: This is the earliest devised method. It shows slower rates of convergence compared to other iterative methods, but it uses very little memory and does not need to solve a matrix system.
- Fast-decoupled-load-flow method is a variation on Newton-Raphson that ex-

ploits the approximate decoupling of active and reactive flows in well-behaved power networks, and additionally fixes the value of the Jacobian during the iteration in order to avoid costly matrix decompositions. Also referred to as "fixed-slope, decoupled NR". Within the algorithm, the Jacobian matrix gets inverted only once, and there are three assumptions. Firstly, the conductance between the buses is zero. Secondly, the magnitude of the bus voltage is one per unit. Thirdly, the sine of phases between buses is zero. Fast decoupled load flow can return the answer within seconds whereas the Newton Raphson method takes much longer. This is useful for real-time management of power grids.

- Holomorphic embedding load flow method: A recently developed method based on advanced techniques of complex analysis. It is direct and guarantees the calculation of the correct (operative) branch, out of the multiple solutions present in the power flow equations.

One-line Diagram

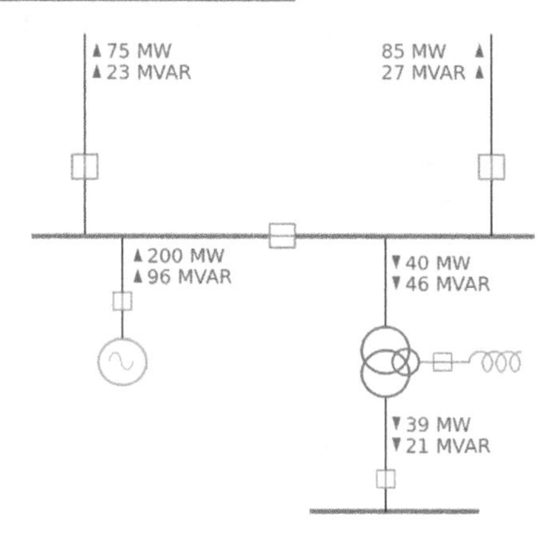

A typical one-line diagram with annotated power flows. Red boxes represent circuit breakers, grey lines represent three-phase bus and interconnecting conductors, the orange circle represents an electric generator, the green spiral is an inductor, and the three overlapping blue circles represent a double-wound transformer with a tertiary winding.

In power engineering, a one-line diagram or single-line diagram (SLD) is a simplified notation for representing a three-phase power system. The one-line diagram has its largest application in power flow studies. Electrical elements such as circuit breakers, transformers, capacitors, bus bars, and conductors are shown by standardized schematic symbols. Instead of representing each of three phases with a separate line or terminal, only one conductor is represented. It is a form of block diagram graphically depicting the paths for power flow between entities of the system. Elements on the diagram do not represent the physical size or location of the electrical equipment, but it is a common convention to organize the diagram with the same left-to-right, top-to-bottom sequence as the switchgear or other apparatus represented. A one-line diagram can also be used to show a high level view of conduit runs for a PLC control system.

Balanced Systems

The theory of three-phase power systems tells us that as long as the loads on each of the three phases are balanced, we can consider each phase separately. In power engineering, this assumption is often useful, and to consider all three phases requires more effort with very little potential advantage. An important and frequent exception is an asymmetric fault on only one or two phases of the system.

A one-line diagram is usually used along with other notational simplifications, such as the per-unit system.

A secondary advantage to using a one-line diagram is that the simpler diagram leaves more space for non-electrical, such as economic, information to be included.

Unbalanced Systems

When using the method of symmetrical components, separate one-line diagrams are made for each of the positive, negative and zero-sequence systems. This simplifies the analysis of unbalanced conditions of a polyphase system. Items that have different impedances for the different phase sequences are identified on the diagrams. For example, in general a generator will have different positive and negative sequence impedance, and certain transformer winding connections block zero-sequence currents. The unbalanced system can be resolved into three single line diagrams for each sequence, and interconnected to show how the unbalanced components add in each part of the system.

References

- McAvinew, Thomas; Mulley, Raymond (2004), Control System Documentation, ISA, p. 165, ISBN 1-55617-896-4
- Guile, A.E.; Paterson, W. (1977), Electrical Power Systems (2nd ed.), Pergamon, p. 4, ISBN 0-08-021729-X
- Tleis, Nasser (2008), Power System Modelling and Fault Analysis, Elsevier, p. 28, ISBN 978-0-7506-8074-5
- Grainger, J.; Stevenson, W. (1994). Power System Analysis. New York: McGraw–Hill. ISBN 0-07-061293-5.

6

Essential Aspects of Power and Energy Engineering

Power and energy engineering concentrates on the branches of electricity generation, efficient energy use and alternative energy sources which form its core aspects. This chapter explores these aspects by elaborating on the history, methods and design of each. The aspects elucidated in this chapter are of vital importance, and provide a better understanding of power and energy engineering.

Electricity Generation

Electricity generation is the process of generating electric power from other sources of primary energy. For electric utilities, it is the first process in the delivery of electricity to consumers. The other processes, electricity transmission, distribution, and electrical power storage and recovery using pumped-storage methods are normally carried out by the electric power industry. Electricity is most often generated at a power station by electromechanical generators, primarily driven by heat engines fuelled by combustion or nuclear fission but also by other means such as the kinetic energy of flowing water and wind. Other energy sources include solar photovoltaics and geothermal power.

Turbo generator

History

The fundamental principles of electricity generation were discovered during the 1820s and early 1830s by the British scientist Michael Faraday. This method is still used today: electricity is generated by the movement of a loop of wire, or disc of copper between the poles of a magnet. Central power stations became economically practical with the development of alternating current power transmission, using power transformers to transmit power at high voltage and with low loss. Electricity has been generated at central stations since 1882. The first power plants were run on water power or coal, and today rely mainly on coal, nuclear, natural gas, hydroelectric, wind generators, and petroleum, with a small amount from solar energy, tidal power, and geothermal sources. The use of power-lines and power-poles have been significantly important in the distribution of electricity.

Methods of Generating Electricity

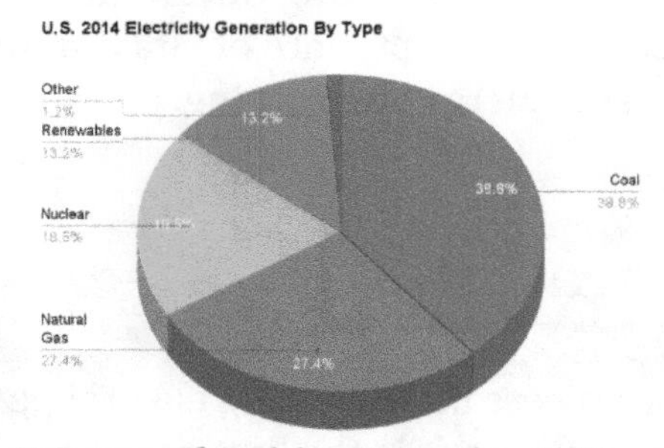

U.S. 2014 Electricity Generation By Type.

There are seven fundamental methods of directly transforming other forms of energy into electrical energy.

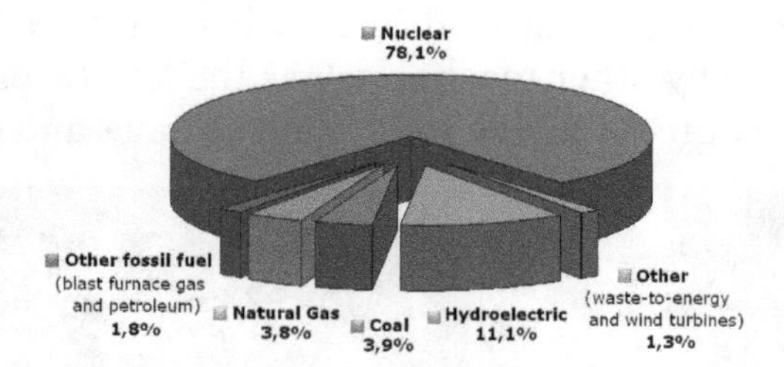

Sources of electricity in France in 2006; nuclear power was the main source.

Static Electricity

Static electricity, from the physical separation and transport of charge (examples: triboelectric effect and lightning). It was the first form discovered and investigated, and the electrostatic generator is still used even in modern devices such as the Van de Graaff generator and MHD generators.

Electromagnetic Induction

In Electromagnetic induction, an electric generator, dynamo or alternator transforms kinetic energy into electricity. This is the most used form for generating electricity and is based on Faraday's law. It can be experimented by rotating a magnet within closed loops of a conducting material (e.g. copper wire). Almost all commercial electrical generation is done using electromagnetic induction, in which mechanical energy forces a generator to rotate.

Turbines

Almost all electrical power on Earth is generated with a turbine, driven by wind, water, steam or burning gas. The turbine drives a generator. There are many different methods of developing mechanical energy, including heat engines, hydro, wind and tidal power. Most electric generation is driven by heat engines. The combustion of fossil fuels supplies most of the heat to these engines, with a significant fraction from nuclear fission and some from renewable sources. The modern steam turbine (invented by Sir Charles Parsons in 1884) currently generates about 80% of the electric power in the world using a variety of heat sources. Power sources include:

- Steam

Large dams such as Three Gorges Dam in China can provide large amounts of hydroelectric power; it has a 22.5 GW capability.

  - Water is boiled by coal burned in a thermal power plant, about 41% of all electricity is generated this way.
  - Nuclear fission heat created in a nuclear reactor creates steam. Less than 15% of electricity is generated this way.
  - Renewables. The steam is generated by Biomass, Solar thermal energy where solar parabolic troughs and solar power towers concentrate sunlight to heat a heat transfer fluid, which is then used to produce steam, or Geothermal power.
  - Natural gas: turbines are driven directly by gases produced by combustion. Combined cycle are driven by both steam and natural gas. They generate power by burning natural gas in a gas turbine and use residual heat to generate steam. At least 20% of the worlds electricity is generated by natural gas.

- Small turbines can be powered by Diesel engines. This is used for back up generation, usually at low voltages. Most large power grids also use diesel generators, originally provided as emergency back up for a specific facility such as a hospital, to feed power into the grid during certain circumstances.
- Water Energy is captured from the movement of water. From falling water, the rise and fall of tides or ocean thermal currents. Each driving a water turbine to produce approximately 16% of the world's electricity. The Perth Wave Energy Project is an early production, submerged buoy, electrical power and direct desalination installation supplying power to HMAS Stirling in Western Australia.
- The windmill was a very early wind turbine. In a solar updraft tower wind is artificially produced. Before 2010 less than 2% of the worlds electricity was produced from wind.

Electrochemistry

Electrochemistry is the direct transformation of chemical energy into electricity, as in a battery. Electrochemical electricity generation is important in portable and mobile applications. Currently, most electrochemical power comes from batteries. Primary cells, such as the common zinc-carbon batteries, act as power sources directly, but many types of cells are used as storage systems rather than primary generation systems. Open electrochemical systems, known as fuel cells, can be used to extract power either from natural fuels or from synthesized fuels. Osmotic power is a possibility at places where salt and fresh water merges.

Large dams such as Hoover Dam can provide large amounts of hydroelectric power; it has 2.07 GW capability.

Photovoltaic Effect

The photovoltaic effect is the transformation of light into electrical energy, as in solar cells. Photovoltaic panels convert sunlight directly to electricity. Although sunlight is free and abundant, solar electricity is still usually more expensive to produce than large-scale mechanically generated power due to the cost of the panels. Low-efficiency silicon solar cells have been decreasing in cost and multijunction cells with close to

30% conversion efficiency are now commercially available. Over 40% efficiency has been demonstrated in experimental systems. Until recently, photovoltaics were most commonly used in remote sites where there is no access to a commercial power grid, or as a supplemental electricity source for individual homes and businesses. Recent advances in manufacturing efficiency and photovoltaic technology, combined with subsidies driven by environmental concerns, have dramatically accelerated the deployment of solar panels. Installed capacity is growing by 40% per year led by increases in Germany, Japan, and the United States.

Thermoelectric Effect

Thermoelectric effect is the direct conversion of temperature differences to electricity, as in thermocouples, thermopiles, and thermionic converters.

A coal-fired power plant in Laughlin, Nevada U.S.A. Owners of this plant ceased operations after declining to invest in pollution control equipment to comply with pollution regulations.

Piezoelectric Effect

The Piezoelectric effect generates electricity from the mechanical strain of electrically anisotropic molecules or crystals. Researchers at the United States Department of Energy's Lawrence Berkeley National Laboratory have developed a piezoelectric generator sufficient to operate a liquid crystal display using thin films of M13 bacteriophage. Piezoelectric devices are used for power generation from mechanical strain, particularly in power harvesting.

Nuclear Transformation

Nuclear transformation is the creation and acceleration of charged particles (examples: betavoltaics or alpha particle emission). The direct conversion of nuclear potential energy to electricity by beta decay is used only on a small scale. In a full-size nuclear power plant, the heat of a nuclear reaction is used to run a heat engine. This drives a generator, which converts mechanical energy into electricity by magnetic induction. Betavoltaics are a type of solid-state power generator which produces electricity from radioactive decay. Fluid-based magnetohydrodynamic (MHD) power generation has been studied as a method for extracting electrical power from nuclear reactors and also from more conventional fuel combustion systems.

Wind turbines usually provide electrical generation in conjunction with other methods of producing power.

Economics of Generation and Production of Electricity

The selection of electricity production modes and their economic viability varies in accordance with demand and region. The economics vary considerably around the world, resulting in widespread selling prices, e.g. the price in Venezuela is 3 cents per kWh while in Denmark it is 40 cents per kWh. Hydroelectric plants, nuclear power plants, thermal power plants and renewable sources have their own pros and cons, and selection is based upon the local power requirement and the fluctuations in demand. All power grids have varying loads on them but the daily minimum is the base load, supplied by plants which run continuously. Nuclear, coal, oil and gas plants can supply base load.

Thermal energy is economical in areas of high industrial density, as the high demand cannot be met by renewable sources. The effect of localized pollution is also minimized as industries are usually located away from residential areas. These plants can also withstand variation in load and consumption by adding more units or temporarily decreasing the production of some units. Nuclear power plants can produce a huge amount of power from a single unit. However, recent disasters in Japan have raised concerns over the safety of nuclear power, and the capital cost of nuclear plants is very high. Hydroelectric power plants are located in areas where the potential energy from falling water can be harnessed for moving turbines and the generation of power. It is not an economically viable source of production where the load varies too much during the annual production cycle and the ability to store the flow of water is limited.

Due to advancements in technology, and with mass production, renewable sources other than hydroelectricity (solar power, wind energy, tidal power, etc.) experienced decreases in cost of production, and the energy is now in many cases cost-comparative with fossil fuels. Many governments around the world provide subsidies to offset the higher cost of any new power production, and to make the installation of renewable energy systems economically feasible. However, their use is frequently limited by their intermittent nature. If natural gas prices are below $3 per million British thermal units, generating electricity from natural gas is cheaper than generating power by burning coal.

Production

The production of electricity in 2009 was 20,053TWh. Sources of electricity were fossil fuels 67%, renewable energy 16% (mainly hydroelectric, wind, solar and biomass), and nuclear power 13%, and other sources were 3%. The majority of fossil fuel usage for the generation of electricity was coal and gas. Oil was 5.5%, as it is the most expensive common commodity used to produce electrical energy. Ninety-two percent of renewable energy was hydroelectric followed by wind at 6% and geothermal at 1.8%. Solar photovoltaic was 0.06%, and solar thermal was 0.004%. Data are from OECD 2011-12 Factbook (2009 data).

Source of Electricity (World total year 2008)							
-	**Coal**	**Oil**	**Natural Gas**	**Nuclear**	**Renewables**	**other**	**Total**
Average electric power (TWh/year)	8,263	1,111	4,301	2,731	3,288	568	20,261
Average electric power (GW)	942.6	126.7	490.7	311.6	375.1	64.8	2311.4
Proportion	41%	5%	21%	13%	16%	3%	100%

data source IEA/OECD

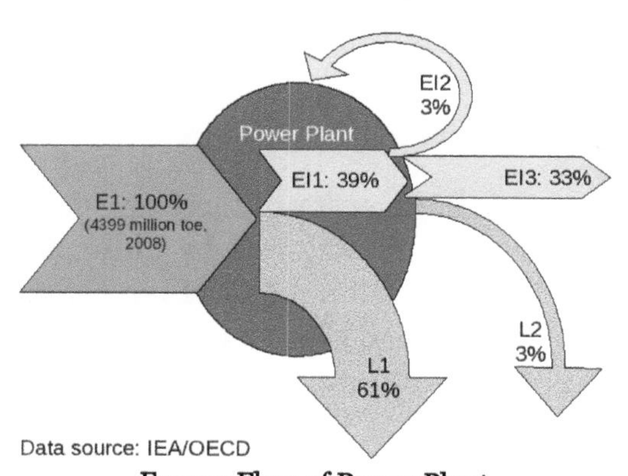

Energy Flow of Power Plant

Total energy consumed at all power plants for the generation of electricity was 4,398,768 ktoe (kilo ton of oil equivalent) which was 36% of the total for primary energy sources (TPES) of 2008.Electricity output (gross) was 1,735,579 ktoe (20,185 TWh), efficiency was 39%, and the balance of 61% was generated heat. A small part (145,141 ktoe, which was 3% of the input total) of the heat was utilized at co-generation heat and power plants. The in-house consumption of electricity and power transmission losses were 289,681 ktoe. The amount supplied to the final consumer was 1,445,285 ktoe (16,430 TWh) which was 33% of the total energy consumed at power plants and heat and power co-generation (CHP) plants.

Historical Results of Production of Electricity

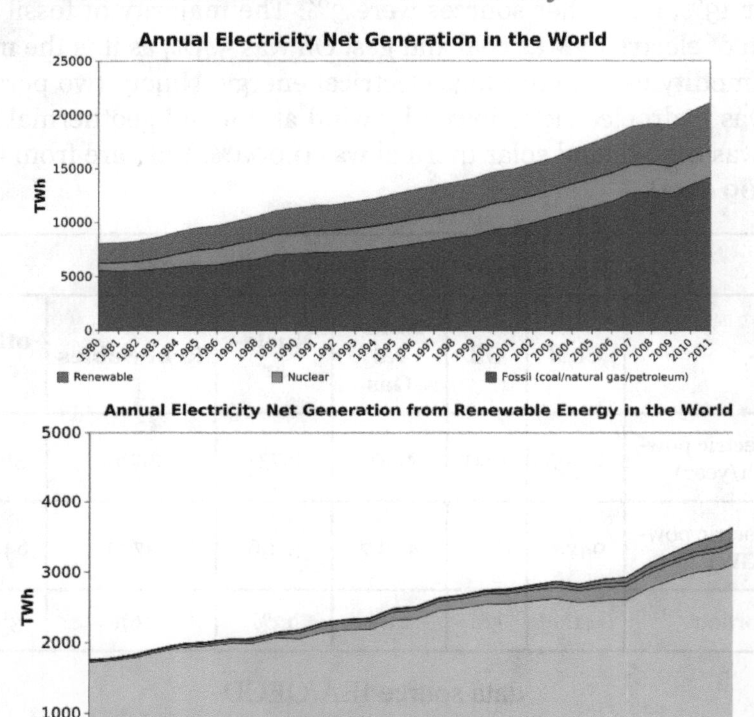

Production by Country

The United States has long been the largest producer and consumer of electricity, with a global share in 2005 of at least 25%, followed by China, Japan, Russia, and India. As of Jan-2010, total electricity generation for the 2 largest generators was as follows: USA: 3992 billion kWh (3992 TWh) and China: 3715 billion kWh (3715 TWh).

Environmental Concerns

Variations between countries generating electrical power affect concerns about the environment. In France only 10% of electricity is generated from fossil fuels, the US is higher at 70% and China is at 80%. The cleanliness of electricity depends on its source. Most scientists agree that emissions of pollutants and greenhouse gases from fossil fuel-based electricity generation account for a significant portion of world greenhouse gas emissions; in the United States, electricity generation accounts for nearly 40% of emissions, the largest of any source. Transportation emissions are close behind, contributing about one-third of U.S. production of carbon dioxide. In the United States, fossil fuel combustion for electric power generation is responsible for 65% of all emissions of sulfur dioxide, the main component of acid rain. Electricity generation is the fourth highest combined source of NOx, carbon monoxide, and particulate matter in the US. In July 2011, the UK parliament tabled a motion that "levels of (carbon) emissions from nuclear power were approximately three times lower per kilowatt hour than those of solar, four times lower than clean coal and 36 times lower than conventional coal".

Lifecycle greenhouse gas emissions by electricity source.

Technology	Description	50th percentile ($g\ CO_2/kWh_e$)
Hydroelectric	reservoir	4
Wind	onshore	12
Nuclear	various generation II reactor types	16
Biomass	various	18
Solar thermal	parabolic trough	22
Geothermal	hot dry rock	45
Solar PV	Polycrystalline silicon	46
Natural gas	various combined cycle turbines without scrubbing	469
Coal	various generator types without scrubbing	1001

Efficient Energy Use

Efficient energy use, sometimes simply called energy efficiency, is the goal to reduce the amount of energy required to provide products and services. For example, insulating a home allows a building to use less heating and cooling energy to achieve and maintain a comfortable temperature. Installing fluorescent lights, LED lights or natural skylights reduces the amount of energy required to attain the same level of illumination compared with using traditional incandescent light bulbs. Improvements in energy efficiency are generally achieved by adopting a more efficient technology or production process or by application of commonly accepted methods to reduce energy losses.

There are many motivations to improve energy efficiency. Reducing energy use reduces energy costs and may result in a financial cost saving to consumers if the energy savings offset any additional costs of implementing an energy efficient technology. Reducing energy use is also seen as a solution to the problem of reducing greenhouse gas emissions. According to the International Energy Agency, improved energy efficiency in buildings, industrial processes and transportation could reduce the world's energy needs in 2050 by one third, and help control global emissions of greenhouse gases.

A spiral-type integrated compact fluorescent lamp, which has been in popular use among North American consumers since its introduction in the mid-1990s.

Energy efficiency and renewable energy are said to be the twin pillars of sustainable energy policy and are high priorities in the sustainable energy hierarchy. In many countries energy efficiency is also seen to have a national security benefit because it can be used to reduce the level of energy imports from foreign countries and may slow down the rate at which domestic energy resources are depleted.

Overview

Energy efficiency has proved to be a cost-effective strategy for building economies without necessarily increasing energy consumption. For example, the state of California began implementing energy-efficiency measures in the mid-1970s, including building code and appliance standards with strict efficiency requirements. During the following years, California's energy consumption has remained approximately flat on a per capita basis while national US consumption doubled. As part of its strategy, California implemented a "loading order" for new energy resources that puts energy efficiency first, renewable electricity supplies second, and new fossil-fired power plants last. States such as Connecticut and New York have created quasi-public Green Banks to help residential and commercial building-owners finance energy efficiency upgrades that reduce emissions and cut consumers' energy costs.

Lovins' Rocky Mountain Institute points out that in industrial settings, "there are abundant opportunities to save 70% to 90% of the energy and cost for lighting, fan, and pump systems; 50% for electric motors; and 60% in areas such as heating, cooling, office equipment, and appliances." In general, up to 75% of the electricity used in the US today could be saved with efficiency measures that cost less than the electricity itself. The same holds true for this is home and there is 78% of electricity uses D in your homehome-owners, leaky ducts have remained an invisible energy culprit for years. In fact, researchers at the US Department of Energy and their consortium, Residential

Energy Efficient Distribution Systems (REEDS) have found that duct efficiency may be as low as 50–70%. The US Department of Energy has stated that there is potential for energy saving in the magnitude of 90 Billion kWh by increasing home energy efficiency.

Other studies have emphasized this. A report published in 2006 by the McKinsey Global Institute, asserted that "there are sufficient economically viable opportunities for energy-productivity improvements that could keep global energy-demand growth at less than 1 percent per annum"—less than half of the 2.2 percent average growth anticipated through 2020 in a business-as-usual scenario. Energy productivity, which measures the output and quality of goods and services per unit of energy input, can come from either reducing the amount of energy required to produce something, or from increasing the quantity or quality of goods and services from the same amount of energy.

The Vienna Climate Change Talks 2007 Report, under the auspices of the United Nations Framework Convention on Climate Change (UNFCCC), clearly shows "that energy efficiency can achieve real emission reductions at low cost."

Appliances

Modern appliances, such as, freezers, ovens, stoves, dishwashers, and clothes washers and dryers, use significantly less energy than older appliances. Installing a clothesline will significantly reduce one's energy consumption as their dryer will be used less. Current energy efficient refrigerators, for example, use 40 percent less energy than conventional models did in 2001. Following this, if all households in Europe changed their more than ten-year-old appliances into new ones, 20 billion kWh of electricity would be saved annually, hence reducing CO_2 emissions by almost 18 billion kg. In the US, the corresponding figures would be 17 billion kWh of electricity and 27,000,000,000 lb (1.2×10^{10} kg) CO_2. According to a 2009 study from McKinsey & Company the replacement of old appliances is one of the most efficient global measures to reduce emissions of greenhouse gases. Modern power management systems also reduce energy usage by idle appliances by turning them off or putting them into a low-energy mode after a certain time. Many countries identify energy-efficient appliances using energy input labeling.

The impact of energy efficiency on peak demand depends on when the appliance is used. For example, an air conditioner uses more energy during the afternoon when it is hot. Therefore, an energy efficient air conditioner will have a larger impact on peak demand than off-peak demand. An energy efficient dishwasher, on the other hand, uses more energy during the late evening when people do their dishes. This appliance may have little to no impact on peak demand.

Building Design

Buildings are an important field for energy efficiency improvements around the world because of their role as a major energy consumer. However, the question of energy use in buildings is not straightforward as the indoor conditions that can be achieved with energy use vary a lot. The measures that keep buildings comfortable, lighting, heating, cooling and ventilation, all consume energy. Typically the level of energy efficiency in a building is measured by dividing energy consumed with the floor area of the building resulting in specific energy consumption (SEC):

Receiving a Gold rating for energy and environmental design in September 2011, the Empire State Building is the tallest and largest LEED certified building in the United States and Western Hemisphere., though it will likely be overtaken by New York's own One World Trade Center.

$$\frac{\textit{Energy consumed}}{\textit{Built area}}$$

However, the issue is more complex as building materials have embodied energy in them. On the other hand, energy can be recovered from the materials when the building is dismantled by reusing materials or burning them for energy. Moreover, the when the building is used, the indoor conditions can vary resulting in higher and lower quality indoor environments. Finally, overall efficiency is affected by the use of the building: is the building occupied most of the time and are spaces efficiently used — or is the building largely empty? It has even been suggested that for a more complete accounting of energy efficiency, SEC should be amended to include these factors:

$$\frac{\textit{Embodied energy} + \textit{Energy consumed} - \textit{Energy recovered}}{\textit{Built area} \times \textit{Utilization rate} \times \textit{Quality factor}}$$

Thus a balanced approach to energy efficiency in buildings should be more comprehensive than simply trying to minimize energy consumed. Issues such as quality of indoor environment and efficiency of space use should be factored in. Thus the measures used to improve energy efficiency can take many different forms. Often they include passive measures that inherently reduce the need to use energy, such as better insulation. Many serve various functions improving the indoor conditions as well as reducing energy use, such as increased use of natural light.

A building's location and surroundings play a key role in regulating its temperature and illumination. For example, trees, landscaping, and hills can provide shade and block wind. In cooler climates, designing northern hemisphere buildings with south facing windows and southern hemisphere buildings with north facing windows increases the amount of sun (ultimately heat energy) entering the building, minimizing energy use, by maximizing passive solar heating. Tight building design, including energy-efficient windows, well-sealed doors, and additional thermal insulation of walls, basement slabs, and foundations can reduce heat loss by 25 to 50 percent.

Dark roofs may become up to 39 °C (70 °F) hotter than the most reflective white surfaces. They transmit some of this additional heat inside the building. US Studies have shown that lightly colored roofs use 40 percent less energy for cooling than buildings with darker roofs. White roof systems save more energy in sunnier climates. Advanced electronic heating and cooling systems can moderate energy consumption and improve the comfort of people in the building.

Proper placement of windows and skylights as well as the use of architectural features that reflect light into a building can reduce the need for artificial lighting. Increased use of natural and task lighting has been shown by one study to increase productivity in schools and offices. Compact fluorescent lights use two-thirds less energy and may last 6 to 10 times longer than incandescent light bulbs. Newer fluorescent lights produce a natural light, and in most applications they are cost effective, despite their higher initial cost, with payback periods as low as a few months.

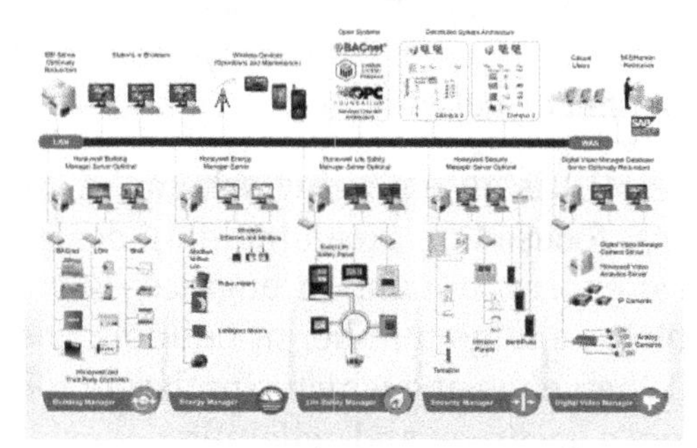

BMS integration

Effective energy-efficient building design can include the use of low cost Passive Infra Reds (PIRs) to switch-off lighting when areas are unnoccupied such as toilets, corridors or even office areas out-of-hours. In addition, lux levels can be monitored using daylight sensors linked to the building's lighting scheme to switch on/off or dim the lighting to pre-defined levels to take into account the natural light and thus reduce consumption. Building Management Systems (BMS) link all of this together in one centralised computer to control the whole building's lighting and power requirements.

In an analysis that integrates a residential bottom-up simulation with an economic multi-sector model, it has been shown that variable heat gains caused by insulation and air-conditioning efficiency can have load-shifting effects that are not uniform on the electricity load. The study also highlighted the impact of higher household efficiency on the power generation capacity choices that are made by the power sector.

The choice of which space heating or cooling technology to use in buildings can have a significant impact on energy use and efficiency. For example, replacing an older 50% efficient natural gas furnace with a new 95% efficient one will dramatically reduce energy use, carbon emissions, and winter natural gas bills. Ground source heat pumps can be even more energy efficient and cost effective. These systems use pumps and compressors to move refrigerant fluid around a thermodynamic cycle in order to "pump" heat against its natural flow from hot to cold, for the purpose of transferring heat into a building from the large thermal reservoir contained within the nearby ground. The end result is that heat pumps typically use four times less electrical energy to deliver an equivalent amount of heat than a direct electrical heater does. Another advantage of a ground source heat pump is that it can be reversed in summertime and operate to cool the air by transferring heat from the building to the ground. The disadvantage of ground source heat pumps is their high initial capital cost, but this is typically recouped within five to ten years as a result of lower energy use.

Smart meters are slowly being adopted by the commercial sector to highlight to staff and for internal monitoring purposes the building's energy usage in a dynamic presentable format. The use of Power Quality Analysers can be introduced into an existing building to assess usage, harmonic distortion, peaks, swells and interruptions amongst others to ultimately make the building more energy-efficient. Often such meters communicate by using wireless sensor networks.

Green Building XML (gbXML) is an emerging schema, a subset of the Building Information Modeling efforts, focused on green building design and operation. gbXML is used as input in several energy simulation engines. But with the development of modern computer technology, a large number of building energy simulation tools are available on the market. When choosing which simulation tool to use in a project, the user must consider the tool's accuracy and reliability, considering the building information they have at hand, which will serve as input for the tool. Yezioro, Dong and Leite developed an artificial intelligence approach towards assessing building performance simulation results and found that more detailed simulation tools have the best simulation performance in terms of heating and cooling electricity consumption within 3% of mean absolute error.

Leadership in Energy and Environmental Design (LEED) is a rating system organized by the US Green Building Council (USGBC) to promote environmental responsibility in building design. They currently offer four levels of certification for existing buildings (LEED-EBOM) and new construction (LEED-NC) based on a building's compliance with the following criteria: Sustainable Sites, Water Efficiency, Energy and Atmosphere, Materials and Resources, Indoor Environmental Quality, and Innovation in Design. In 2013, USGBC developed the LEED Dynamic Plaque, a tool to track building performance against LEED metrics and a potential path to recertification. The following year, the council collaborated with Honeywell to pull data on energy and water use, as well as indoor air quality from a BAS to automatically update the plaque, providing a near-real-time view of performance. The USGBC office in Washington, D.C. is one of the first buildings to feature the live-updating LEED Dynamic Plaque.

A deep energy retrofit is a whole-building analysis and construction process that uses to achieve much larger energy savings than conventional energy retrofits. Deep energy retrofits can be applied to both residential and non-residential ("commercial") buildings. A deep energy retrofit typically results in energy savings of 30 percent or more, perhaps spread over several years, and may significantly improve the building value. The Empire State Building has undergone a deep energy retrofit process that was completed in 2013. The project team, consisting of representatives from Johnson Controls, Rocky Mountain Institute, Clinton Climate Initiative, and Jones Lang LaSalle will have achieved an annual energy use reduction of 38% and $4.4 million. For example, the 6,500 windows were remanufactured onsite into superwindows which block heat but pass light. Air conditioning operating costs on hot days were reduced and this saved $17 million of the project's capital cost immediately, partly funding other retrofitting. Receiving a gold Leadership in Energy and Environmental Design (LEED) rating in September 2011, the Empire State Building is the tallest LEED certified building in the United States. The Indianapolis City-County Building recently underwent a deep energy retrofit process, which has achieved an annual energy reduction of 46% and $750,000 annual energy saving.

Energy retrofits, including deep, and other types undertaken in residential, commercial or industrial locations are generally supported through various forms of financing or incentives. Incentives include pre-packaged rebates where the buyer/user may not even be aware that the item being used has been rebated or "bought down". "Upstream" or "Midstream" buy downs are common for efficient lighting products. Other rebates are more explicit and transparent to the end user through the use of formal applications. In addition to rebates, which may be offered through government or utility programs, governments sometimes offer tax incentives for energy efficiency projects. Some entities offer rebate and payment guidance and facilitation services that enable energy end use customers tap into rebate and incentive programs.

To evaluate the economic soundness of energy efficiency investments in buildings, cost-effectiveness analysis or CEA can be used. A CEA calculation will produce the value of energy saved, sometimes called negawatts, in $/kWh. The energy in such a calculation is virtual in the sense that it was never consumed but rather saved due to some energy efficiency investment being made. Thus CEA allows comparing the price of negawatts with price of energy such as electricity from the grid or the cheapest renewable alternative. The benefit of the CEA approach in energy systems is that it avoids the need to guess future energy prices for the purposes of the calculation, thus removing the major source of uncertainty in the appraisal of energy efficiency investments.

Energy Efficiency by Country

Europe

Energy efficiency targets for 2020 and 2030.

The EU has set itself a 20% energy savings target by 2020 when compared to the projected use of energy in 2020 – roughly equivalent to turning off 400 power stations. At an EU summit in October 2014, EU countries agreed on a new energy efficiency target of 27% or greater by 2030. One mechanism used to achieve the target of 27% is the 'Suppliers Obligations & White Certificates'.

Australia

The Australian national government is actively leading the country in efforts to increase their energy efficiency, mainly through the government's Department of Industry and Science. In July 2009, the Council of Australian Governments, which represents the individual states and territories of Australia, agreed to a National Strategy on Energy Efficiency (NSEE).

This is a ten-year plan accelerate the implementation of a nationwide adoption of energy efficient practices and a preparation for the country's transformation into a low carbon future. There are several different areas of energy use addressed within the NSEE. But, the chapter devoted to the approach on energy efficiency that is to be adopted on a national level stresses four points in achieving stated levels of energy efficiency. They are:

- To help households and businesses transition to a low carbon future
- To streamline the adoption of efficient energy
- To make buildings more energy efficient
- For governments to work in partnership and lead the way to energy efficiency

The overriding agreement that governs this strategy is the National Partnership Agreement on Energy Efficiency.

This document also explains the role of both the commonwealth and the individual states and territories in the NSEE, as well provides for the creation of benchmarks and measurement devices which will transparently show the nation's progress in relation to the stated goals, and addresses the need for funding of the strategy in order to enable it to move forward.

Germany

Energy efficiency is central to energy policy in Germany. As of late 2015, national policy includes the following efficiency and consumption targets (with actual values for

2014):

Efficiency and consumption target	2014	2020	2050
Primary energy consumption (base year 2008)	−8.7%	−20%	−50%
Final energy productivity (2008–2050)	1.6%/year (2008–2014)	2.1%/year (2008–2050)	
Gross electricity consumption (base year 2008)	−4.6%	−10%	−25%
Primary energy consumption in buildings (base year 2008)	−14.8%		−80%
Heat consumption in buildings (base year 2008)	−12.4%	−20%	
Final energy consumption in transport (base year 2005)	1.7%	−10%	−40%

Recent progress toward improved efficiency has been steady aside from the financial crisis of 2007–2008. Some however believe energy efficiency is still under-recognised in terms of its contribution to Germany's energy transformation (or Energiewende).

Efforts to reduce final energy consumption in transport sector have not been successful, with a growth of 1.7% between 2005–2014. This growth is due to both road passenger and road freight transport. Both sectors increased their overall distance travelled to record the highest figures ever for Germany. Rebound effects played a significant role, both between improved vehicle efficiency and the distance travelled, and between improved vehicle efficiency and an increase in vehicle weights and engine power.

On 3 December 2014, the German federal government released its National Action Plan on Energy Efficiency (NAPE). The areas covered are the energy efficiency of buildings, energy conservation for companies, consumer energy efficiency, and transport energy efficiency. The policy contains both immediate and forward-looking measures. The central short-term measures of NAPE include the introduction of competitive tendering for energy efficiency, the raising of funding for building renovation, the introduction of tax incentives for efficiency measures in the building sector, and the setting up energy efficiency networks together with business and industry. German industry is expected to make a sizeable contribution.

On 12 August 2016, the German government released a green paper on energy efficiency for public consultation (in German). It outlines the potential challenges and actions needed to reduce energy consumption in Germany over the coming decades. At the document's launch, economics and energy minister Sigmar Gabriel said "we do not need to produce, store, transmit and pay for the energy that we save". The green paper prioritizes the efficient use of energy as the "first" response and also outlines opportunities for sector coupling, including using renewable power for heating and transport. Other proposals include a flexible energy tax which rises as petrol prices fall, thereby incentivizing fuel conservation despite low oil prices.

United States

A 2011 Energy Modeling Forum study covering the United States examines how energy efficiency opportunities will shape future fuel and electricity demand over the next several decades. The US economy is already set to lower its energy and carbon intensity, but explicit policies will be necessary to meet climate goals. These policies include: a carbon tax, mandated standards for more efficient appliances, buildings and vehicles, and subsidies or reductions in the upfront costs of new more energy efficient equipment.

Industry

Industries use a large amount of energy to power a diverse range of manufacturing and resource extraction processes. Many industrial processes require large amounts of heat and mechanical power, most of which is delivered as natural gas, petroleum fuels and as electricity. In addition some industries generate fuel from waste products that can be used to provide additional energy.

Because industrial processes are so diverse it is impossible to describe the multitude of possible opportunities for energy efficiency in industry. Many depend on the specific technologies and processes in use at each industrial facility. There are, however, a number of processes and energy services that are widely used in many industries.

Various industries generate steam and electricity for subsequent use within their facilities. When electricity is generated, the heat that is produced as a by-product can be captured and used for process steam, heating or other industrial purposes. Conventional electricity generation is about 30% efficient, whereas combined heat and power (also called co-generation) converts up to 90 percent of the fuel into usable energy.

Advanced boilers and furnaces can operate at higher temperatures while burning less fuel. These technologies are more efficient and produce fewer pollutants.

Over 45 percent of the fuel used by US manufacturers is burnt to make steam. The typical industrial facility can reduce this energy usage 20 percent (according to the US Department of Energy) by insulating steam and condensate return lines, stopping steam leakage, and maintaining steam traps.

Electric motors usually run at a constant speed, but a variable speed drive allows the motor's energy output to match the required load. This achieves energy savings ranging from 3 to 60 percent, depending on how the motor is used. Motor coils made of superconducting materials can also reduce energy losses. Motors may also benefit from voltage optimisation.

Industry uses a large number of pumps and compressors of all shapes and sizes and in a wide variety of applications. The efficiency of pumps and compressors depends on many factors but often improvements can be made by implementing better process control and better maintenance practices. Compressors are commonly used to provide compressed air which is used for sand blasting, painting, and other power tools. According to the US Department of Energy, optimizing compressed air systems by installing variable speed drives, along with preventive maintenance to detect and fix air leaks, can improve energy efficiency 20 to 50 percent.

Transportation

Automobiles

The estimated energy efficiency for an automobile is 280 Passenger-Mile/10^6 Btu. There are several ways to enhance a vehicle's energy efficiency. Using improved aerodynamics to minimize drag can increase vehicle fuel efficiency. Reducing vehicle weight can also improve fuel economy, which is why composite materials are widely used in car bodies.

Toyota Prius used by NYPD Traffic Enforcement

More advanced tires, with decreased tire to road friction and rolling resistance, can save gasoline. Fuel economy can be improved by up to 3.3% by keeping tires inflated to the correct pressure. Replacing a clogged air filter can improve a cars fuel consumption by as much as 10 percent on older vehicles. On newer vehicles (1980s and up) with fuel-injected, computer-controlled engines, a clogged air filter has no effect on mpg but replacing it may improve acceleration by 6-11 percent.

Turbochargers can increase fuel efficiency by allowing a smaller displacement engine. The 'Engine of the year 2011' is a Fiat 500 engine equipped with an MHI turbocharger. "Compared with a 1.2-liter 8v engine, the new 85 HP turbo has 23% more power and a 30% better performance index. The performance of the two-cylinder is not only equivalent to a 1.4-liter 16v engine, but fuel consumption is 30% lower."

Energy-efficient vehicles may reach twice the fuel efficiency of the average automobile. Cutting-edge designs, such as the diesel Mercedes-Benz Bionic concept vehicle have achieved a fuel efficiency as high as 84 miles per US gallon (2.8 L/100 km; 101 mpg_{-imp}), four times the current conventional automotive average.

The mainstream trend in automotive efficiency is the rise of electric vehicles (all@electric or hybrid electric). Hybrids, like the Toyota Prius, use regenerative braking to recapture energy that would dissipate in normal cars; the effect is especially pronounced in city driving. Plug-in hybrids also have increased battery capacity, which makes it possible to drive for limited distances without burning any gasoline; in this case, energy efficiency is dictated by whatever process (such as coal-burning, hydroelectric, or renewable source) created the power. Plug-ins can typically drive for around 40 miles (64 km) purely on electricity without recharging; if the battery runs low, a gas engine kicks in allowing for extended range. Finally, all-electric cars are also growing in popularity; the Tesla Model S sedan is the only high-performance all-electric car currently on the market.

Street Lighting

Cities around the globe light up millions of streets with 300 million lights. Some cities are seeking to reduce street light power consumption by dimming lights during off-peak hours or switching to LED lamps. It is not clear whether the high luminous efficiency of LEDs will lead to real reductions in energy, as cities may end up installing extra lamps or lighting areas more brightly than in the past.

Aircraft

There are several ways to reduce energy usage in air transportation, from modifications to the planes themselves, to how air traffic is managed. As in cars, turbochargers are an effective way to reduce energy consumption; however, instead of allowing for the use of a smaller-displacement engine, turbochargers in jet turbines operate by compressing the thinner air at higher altitudes. This allows the engine to operate as if it were at sea-level pressures while taking advantage of the reduced drag on the aircraft at higher altitudes.

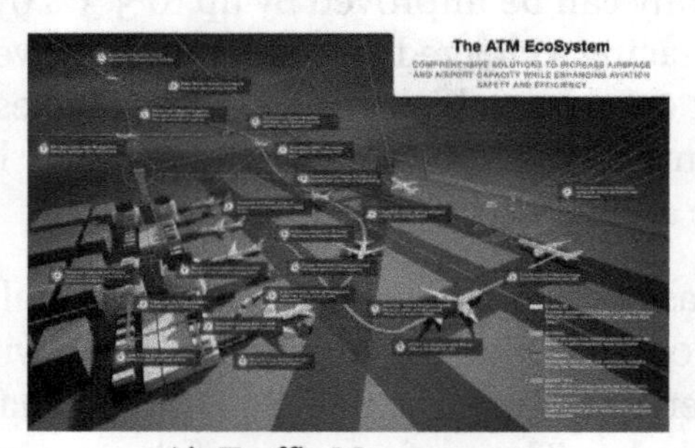

Air Traffic Management

Air traffic management systems are another way to increase the efficiency of not just the aircraft but the airline industry as a whole. New technology allows for superior automation of takeoff, landing, and collision avoidance, as well as within airports, from simple things like HVAC and lighting to more complex tasks such as security and scanning.

Alternative Fuels

Alternative fuels, known as non-conventional or advanced fuels, are any materials or substances that can be used as fuels, other than conventional fuels. Some well known alternative fuels include biodiesel, bioalcohol (methanol, ethanol, butanol), chemically stored electricity (batteries and fuel cells), hydrogen, non-fossil methane, non-fossil natural gas, vegetable oil, and other biomass sources.

Typical Brazilian filling station with four alternative fuels for sale: biodiesel (B3), gasohol (E25), neat ethanol (E100), and compressed natural gas (CNG). Piracicaba, Brazil.

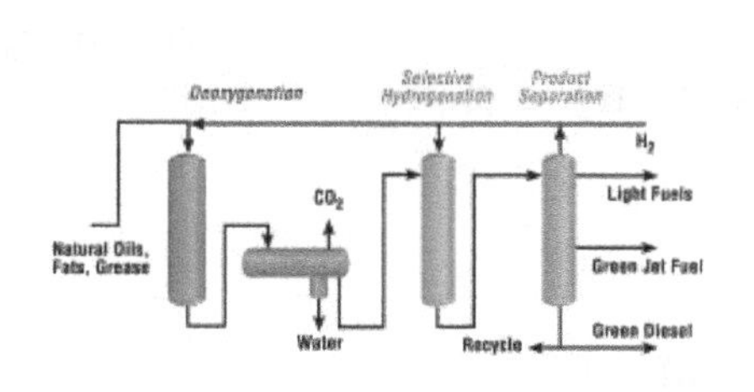

Green jet fuel production

Green diesel production

Energy Conservation

Energy conservation is broader than energy efficiency in including active efforts to decrease energy consumption, for example through behaviour change, in addition to using energy more efficiently. Examples of conservation without efficiency improvements are heating a room less in winter, using the car less, air-drying your clothes instead of using the dryer, or enabling energy saving modes on a computer. As with other definitions, the boundary between efficient energy use and energy conservation can be fuzzy, but both are important in environmental and economic terms. This is especially the case when actions are directed at the saving of fossil fuels. Energy conservation is a challenge requiring policy programmes, technological development and behavior change to go hand in hand. Many energy intermediary organisations, for example governmental or non-governmental organisations on local, regional, or national level, are working on often publicly funded programmes or projects to meet this challenge. Psychologists have also engaged with the issue of energy conservation and have provided guidelines for realizing behavior change to reduce energy consumption while taking technological and policy considerations into account.

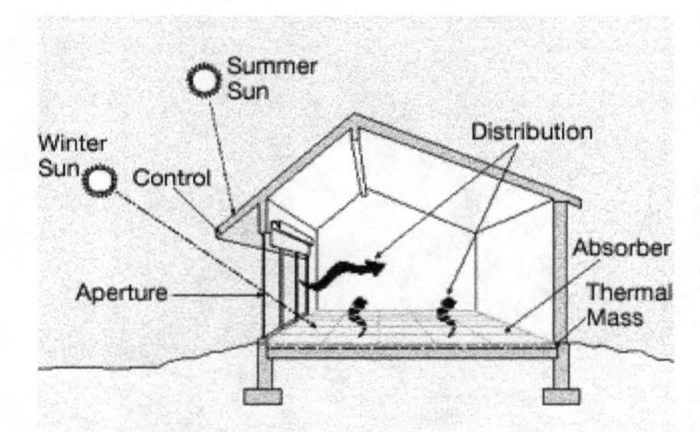

Elements of passive solar energy design, shown in a direct gain application

The National Renewable Energy Laboratory maintains a comprehensive list of apps useful for energy efficiency.

Commercial property managers that plan and manage energy efficiency projects generally use a software platform to perform energy audits and to collaborate with contractors to understand their full range of options. The Department of Energy (DOE) Software Directory describes EnergyActio software, a cloud based platform designed for this purpose.

Sustainable Energy

Energy efficiency and renewable energy are said to be the "twin pillars" of a sustainable energy policy. Both strategies must be developed concurrently in order to stabilize and reduce carbon dioxide emissions. Efficient energy use is essential to slowing the energy demand growth so that rising clean energy supplies can make deep cuts in fossil fuel use. If energy use grows too rapidly, renewable energy development will chase a receding target. Likewise, unless clean energy supplies come online rapidly, slowing demand growth will only begin to reduce total carbon emissions; a reduction in the carbon content of energy sources is also needed. A sustainable energy economy thus requires major commitments to both efficiency and renewables.

Rebound Effect

If the demand for energy services remains constant, improving energy efficiency will reduce energy consumption and carbon emissions. However, many efficiency improvements do not reduce energy consumption by the amount predicted by simple engineering models. This is because they make energy services cheaper, and so consumption of those services increases. For example, since fuel efficient vehicles make travel cheaper, consumers may choose to drive farther, thereby offsetting some of the potential energy savings. Similarly, an extensive historical analysis of technological efficiency improvements has conclusively shown that energy efficiency improvements were almost always outpaced by economic growth, resulting in a net increase in resource use and associated pollution. These are examples of the direct rebound effect.

Estimates of the size of the rebound effect range from roughly 5% to 40%. The rebound effect is likely to be less than 30% at the household level and may be closer to 10% for transport. A rebound effect of 30% implies that improvements in energy efficiency should achieve 70% of the reduction in energy consumption projected using engineering models. The rebound effect may be particularly large for lighting, because in contrast to tasks like transport there is effectively no upper limit on how much light could be considered useful. In fact, it appears that lighting has accounted for about 0.7% of GDP across many societies and hundreds of years, implying a rebound effect of 100%.

Organisations and Programs

International

- 80 Plus
- 2000-watt society
- IEA Solar Heating & Cooling Implementing Agreement Task 13
- International Institute for Energy Conservation
- International Energy Agency (e.g. One Watt initiative)
- International Electrotechnical Commission
- International Partnership for Energy Efficiency Cooperation
- World Sustainable Energy Days

China

- National Development and Reform Commission
- National Energy Conservation Center
- Energy Research Institute, NDRC

Australia

- Department of Climate Change and Energy Efficiency
- Department of the Environment, Water, Heritage and the Arts
- Sustainable House Day

European Union

- Building energy rating
- Eco-Design of Energy-Using Products Directive
- Energy efficiency in Europe
- Orgalime, the European engineering industries association

Iceland

- Marorka
- Africa

India

- 88888 Lights Out
- Bureau of Energy Efficiency
- Energy Efficiency Services Limited

Japan

- Cool Biz campaign

Lebanon

- The Lebanese Center for Energy Conservation

United Kingdom

- The Carbon Trust
- Energy Saving Trust
- National Energy Action
- National Energy Foundation
- Creative Energy Homes
- Energy Managers Association

United States

- Alliance to Save Energy
- American Council for an Energy-Efficient Economy (ACEEE)
- Building Codes Assistance Project

- Building Energy Codes Program
- Consortium for Energy Efficiency
- Energy Star, from United States Environmental Protection Agency
- Enervee
- Industrial Assessment Center
- National Electrical Manufacturers Association
- Rocky Mountain Institute
- Indian energy strategies

Alternative Energy

Alternative energy is any energy source that is an alternative to fossil fuel. These alternatives are intended to address concerns about such fossil fuels, such as its high carbon dioxide emissions, an important factor in global warming. Ocean, hydroelectric, wind, geothermal and solar power are all alternative sources of energy.

Offshore wind turbines near Copenhagen

The nature of what constitutes an alternative energy source has changed considerably over time, as have controversies regarding energy use. Because of the variety of energy choices and differing goals of their advocates, defining some energy types as "alternative" is considered controversial.

Definitions

Source	Definition
Oxford Dictionary	Energy fueled into ways that do not use up natural resources or harm the environment.
Princeton WordNet	Energy derived from sources that do not use up natural resources or harm the environment.

Responding to Climate Change 2007	Energy derived from nontraditional sources (e.g., compressed natural gas, solar, hydroelectric, wind).
Natural Resources Defense Council	Energy that is not popularly used and is usually environmentally sound, such as solar or wind energy (as opposed to fossil fuels).
Materials Management	Fuel sources that are other than those derived from fossil fuels. Typically used interchangeably for renewable energy. Examples include: wind, solar, biomass, wave and tidal energy.
Torridge District Council	Energy generated from alternatives to fossil fuels. Need not be renewable.

History

Historians of economies have examined the key transitions to alternative energies and regard the transitions as pivotal in bringing about significant economic change. Prior to the shift to an alternative energy, supplies of the dominant energy type became erratic, accompanied by rapid increases in energy prices.

Coal as an Alternative to Wood

In the late medieval period, coal was the new alternative fuel to save the society from overuse of the dominant fuel, wood. The deforestation had resulted in shortage of wood, at that time soft coal appeared as a savior. Historian Norman F. Cantor describes how:

> "Europeans had lived in the midst of vast forests throughout the earlier medieval centuries. After 1250 they became so skilled at deforestation that by 1500 AD they were running short of wood for heating and cooking... By 1500 Europe was on the edge of a fuel and nutritional disaster, [from] which it was saved in the sixteenth century only by the burning of soft coal and the cultivation of potatoes and maize. "

Petroleum as an Alternative to Whale Oil

Whale oil was the dominant form of lubrication and fuel for lamps in the early 19th century, but the depletion of the whale stocks by mid century caused whale oil prices to skyrocket setting the stage for the adoption of petroleum which was first commercialized in Pennsylvania in 1859.

Ethanol as an Alternative to Fossil Fuels

In 1917, Alexander Graham Bell advocated ethanol from corn, wheat and other foods as an alternative to coal and oil, stating that the world was in measurable distance of depleting these fuels. For Bell, the problem requiring an alternative was lack of renewability of orthodox energy sources. Since the 1970s, Brazil has had an ethanol fuel program which has allowed the country to become the world's second largest producer of ethanol (after the United States) and the world's largest exporter. Brazil's ethanol fuel program uses modern equipment and cheap sugar cane as feedstock, and the residual cane-waste (bagasse) is used to process heat and power. There are no longer light vehicles in Brazil running on pure gasoline. By the end of 2008 there were 35,000 filling stations throughout Brazil with at least one ethanol pump.

Cellulosic ethanol can be produced from a diverse array of feedstocks, and involves the use of the whole crop. This new approach should increase yields and reduce the carbon footprint because the amount of energy-intensive fertilizers and fungicides will remain the same, for a higher output of usable material. As of 2008, there are nine commercial cellulosic ethanol plants which are either operating, or under construction, in the United States.

Second-generation biofuels technologies are able to manufacture biofuels from inedible biomass and could hence prevent conversion of food into fuel." As of July 2010, there is one commercial second-generation (2G) ethanol plant Inbicon Biomass Refinery, which is operating in Denmark.

Coal Gasification as an Alternative to Petroleum

In the 1970s, President Jimmy Carter's administration advocated coal gasification as an alternative to expensive imported oil. The program, including the Synthetic Fuels Corporation was scrapped when petroleum prices plummeted in the 1980s. The carbon footprint and environmental impact of coal gasification are both very high.

Common Types of Alternative Energy

- Hydro electricity captures energy from falling water.
- Nuclear energy uses nuclear fission to release energy stored in the atomic bonds of heavy elements.
- Wind energy is the generation of electricity from wind, commonly by using propeller-like turbines.
- Solar energy is the use of sunlight. Light can be changed into thermal (heat) energy and electric energy.
- Geothermal energy is the use of the earth's internal heat to boil water for heating buildings or generating electricity.
- Biofuel and Ethanol are plant-derived gasoline substitutes for powering vehicles.
- Hydrogen can serve as a means of delivering energy produced by various technologies.

Enabling Technologies

Heat pumps and Thermal energy storage are technologies which use energy sources that are normally cost-prohibitive. Also, heat pumps have the advantage of leveraging electrical power (or in some cases mechanical or thermal power) by using it to extract additional energy from a low-energy-density source (such as sea or lake water, the ground or the air).

Thermal storage technologies allow heat or cold to be stored for periods of time ranging from diurnal to interseasonal, and can involve storage of sensible energy (i.e. by changing the temperature of a medium) or latent energy (e.g. through phase changes of a medium (i.e. changes from solid to liquid or vice versa), such as between water and slush or ice). Energy sources can be natural (via solar-thermal collectors, or dry cooling towers used to collect winter's cold), waste energy (such as from HVAC equipment, industrial processes or power plants), or surplus energy (such as seasonally from hydropower projects or intermittently from wind farms). The Drake Landing Solar Community (Alberta, Canada) is illustrative. Borehole thermal energy storage allows the community to get 97% of its year-round heat from solar collectors on the garage roofs, which most of the heat collected in summer. The storages can be insulated tanks, borehole clusters in substrates ranging from gravel to bedrock, deep aquifers, or shallow pits that are lined and insulated. Some applications require inclusion of a heat pump.

Renewable Energy vs Non-renewable Energy

Renewable energy is generated from natural resources—such as sunlight, wind, rain, tides and geothermal heat—which are renewable (naturally replenished). When comparing the processes for producing energy, there remain several fundamental differences between renewable energy and fossil fuels. The process of producing oil, coal, or natural gas fuel is a difficult and demanding process that requires a great deal of complex equipment, physical and chemical processes. On the other hand, alternative energy can be widely produced with basic equipment and natural processes. Wood, the most renewable and available alternative energy, burns the same amount of carbon it would emit if it degraded naturally. Nuclear power is an alternative to fossil fuels that is non-renewable, like fossil fuels, nuclear ones are a finite resource.

Ecologically Friendly Alternatives

Renewable energy sources such as burning biomass are sometimes regarded as a carbon-neutral alternative to fossil fuels. Renewables are not inherently alternative energies for this purpose. For example, the Netherlands, once leader in use of palm oil as a biofuel, has suspended all subsidies for palm oil due to the scientific evidence that their use "may sometimes create more environmental harm than fossil fuels". The Netherlands government and environmental groups are trying to trace the origins of imported palm oil, to certify which operations produce the oil in a responsible manner. Regarding biofuels from foodstuffs, the realization that converting the entire grain harvest of the US would only produce 16% of its auto fuel needs, and the decimation of Brazil's

CO_2 absorbing tropical rain forests to make way for biofuel production has made it clear that placing energy markets in competition with food markets results in higher food prices and insignificant or negative impact on energy issues such as global warming or dependence on foreign energy. Recently, alternatives to such undesirable sustainable fuels are being sought, such as commercially viable sources of cellulosic ethanol.

Relatively New Concepts for Alternative Energy

Carbon-neutral and Negative Fuels

Carbon-neutral fuels are synthetic fuels (including methane, gasoline, diesel fuel, jet fuel or ammonia) produced by hydrogenating waste carbon dioxide recycled from power plant flue-gas emissions, recovered from automotive exhaust gas, or derived from carbonic acid in seawater. Commercial fuel synthesis companies suggest they can produce synthetic fuels for less than petroleum fuels when oil costs more than $55 per barrel. Renewable methanol (RM) is a fuel produced from hydrogen and carbon dioxide by catalytic hydrogenation where the hydrogen has been obtained from water electrolysis. It can be blended into transportation fuel or processed as a chemical feedstock.

The George Olah carbon dioxide recycling plant operated by Carbon Recycling International in Grindavík, Iceland has been producing 2 million liters of methanol transportation fuel per year from flue exhaust of the Svartsengi Power Station since 2011. It has the capacity to produce 5 million liters per year. A 250 kilowatt methane synthesis plant was constructed by the Center for Solar Energy and Hydrogen Research (ZSW) at Baden-Württemberg and the Fraunhofer Society in Germany and began operating in 2010. It is being upgraded to 10 megawatts, scheduled for completion in autumn, 2012. Audi has constructed a carbon-neutral liquefied natural gas (LNG) plant in Werlte, Germany. The plant is intended to produce transportation fuel to offset LNG used in their A3 Sportback g-tron automobiles, and can keep 2,800 metric tons of CO_2 out of the environment per year at its initial capacity. Other commercial developments are taking place in Columbia, South Carolina, Camarillo, California, and Darlington, England.

Such fuels are considered carbon-neutral because they do not result in a net increase in atmospheric greenhouse gases. To the extent that synthetic fuels displace fossil fuels, or if they are produced from waste carbon or seawater carbonic acid, and their combustion is subject to carbon capture at the flue or exhaust pipe, they result in negative carbon dioxide emission and net carbon dioxide removal from the atmosphere, and thus constitute a form of greenhouse gas remediation.

Such renewable fuels alleviate the costs and dependency issues of imported fossil fuels without requiring either electrification of the vehicle fleet or conversion to hydrogen or other fuels, enabling continued compatible and affordable vehicles. Carbon-neutral fuels offer relatively low cost energy storage, alleviating the problems of wind and solar intermittency, and they enable distribution of wind, water, and solar power through existing natural gas pipelines.

Nighttime wind power is considered the most economical form of electrical power with which to synthesize fuel, because the load curve for electricity peaks sharply during the warmest hours of the day, but wind tends to blow slightly more at night than during the day, so, the price of nighttime wind power is often much less expensive than any alternative. Germany has built a 250 kilowatt synthetic methane plant which they are scaling up to 10 megawatts.

Algae Fuel

Algae fuel is a biofuel which is derived from algae. During photosynthesis, algae and other photosynthetic organisms capture carbon dioxide and sunlight and convert it into oxygen and biomass. This is usually done by placing the algae between two panes of glass. The algae creates three forms of energy fuel: heat (from its growth cycle), biofuel (the natural "oil" derived from the algae), and biomass (from the algae itself, as it is harvested upon maturity).

The heat can be used to power building systems (such as heat process water) or to produce energy. Biofuel is oil extracted from the algae upon maturity, and used to create energy similar to the use of biodiesel. The biomass is the matter left over after extracting the oil and water, and can be harvested to produce combustible methane for energy production, similar to the warmth felt in a compost pile or the methane collected from biodegradable materials in a landfill. Additionally, the benefits of algae biofuel are that it can be produced industrially, as well as vertically (i.e. as a building facade), thereby obviating the use of arable land and food crops (such as soy, palm, and canola).

Biomass Briquettes

Biomass briquettes are being developed in the developing world as an alternative to charcoal. The technique involves the conversion of almost any plant matter into compressed briquettes that typically have about 70% the calorific value of charcoal. There are relatively few examples of large scale briquette production. One exception is in North Kivu, in eastern Democratic Republic of Congo, where forest clearance for charcoal production is considered to be the biggest threat to Mountain Gorilla habitat. The staff of Virunga National Park have successfully trained and equipped over 3500 people to produce biomass briquettes, thereby replacing charcoal produced illegally inside the national park, and creating significant employment for people living in extreme poverty in conflict affected areas.

Biogas Digestion

Biogas digestion deals with harnessing the methane gas that is released when waste breaks down. This gas can be retrieved from garbage or sewage systems. Biogas digesters are used to process methane gas by having bacteria break down biomass in an anaerobic environment. The methane gas that is collected and refined can be used as an energy source for various products.

Biological Hydrogen Production

Hydrogen gas is a completely clean burning fuel; its only by-product is water. It also contains relatively high amount of energy compared with other fuels due to its chemical structure.

$$2H_2 + O_2 \rightarrow 2H_2O + \text{High Energy}$$

$$\text{High Energy} + 2H_2O \rightarrow 2H_2 + O_2$$

This requires a high-energy input, making commercial hydrogen very inefficient. Use of a biological vector as a means to split water, and therefore produce hydrogen gas, would allow for the only energy input to be solar radiation. Biological vectors can include bacteria or more commonly algae. This process is known as biological hydrogen production. It requires the use of single celled organisms to create hydrogen gas through fermentation. Without the presence of oxygen, also known as an anaerobic environment, regular cellular respiration cannot take place and a process known as fermentation takes over. A major by-product of this process is hydrogen gas. If we could implement this on a large scale, then we could take sunlight, nutrients and water and create hydrogen gas to be used as a dense source of energy. Large-scale production has proven difficult. It was not until 1999 that we were able to even induce these anaerobic conditions by sulfur deprivation. Since the fermentation process is an evolutionary back up, turned on during stress, the cells would die after a few days. In 2000, a two-stage process was developed to take the cells in and out of anaerobic conditions and therefore keep them alive. For the last ten years, finding a way to do this on a large-scale has been the main goal of research. Careful work is being done to ensure an efficient process before large-scale production, however once a mechanism is developed, this type of production could solve our energy needs.

Hydroelectricity

Hydroelectricity provided 75% of the worlds renewable electricity in 2013. Much of the electricity we use today is a result of the heyday of conventional hydroelectric development between 1960 and 1980. According to a report published in 2000 the 1990s dam construction has virtually ceased in Europe and North America due to environmental concerns. Globally there is a trend towards more hydroelectricity. From 2004 to 2014 the installed capacity rose from 715 to 1,055 GW. A popular alternative to the large dams of the past is run-of-the-river where there is no water stored behind a dam and generation usually varies with seasonal rainfall. Using run-of-the-river in wet seasons and solar in dry seasons can balance seasonal variations for both.

In recent years, we discovered the advantage from falling water. The theory of hydroelectric is to build a dam on a large river that has a large drop in elevation .The dam store lots of water behind it in the reservoir. Near the bottom of the dam wall, there is the water intake. Gravity causes it to fall through the penstock inside the dam. At the end of the penstock, there is a turbine propeller, which is turned by the moving water. The shaft from the turbine connected to the generator, which produces the power. Power lines are connected to the generator that carry electricity to your home and mine. There are two advantages of hydroelectric power. To begin, low capital cost to produce energy because the water is an initial substance. Next, it emits less greenhouse gases since water is green chemical.

Offshore Wind

Offshore wind farms are similar to regular wind farms, but are located in the ocean. Offshore wind farms can be placed in water up to 40 metres (130 ft) deep, whereas floating wind turbines can float in water up to 700 metres (2,300 ft) deep. The advantage of having a floating wind farm is to be able to harness the winds from the open ocean. Without any obstructions such as hills, trees and buildings, winds from the open ocean can reach up to speeds twice as fast as coastal areas.

Significant generation of offshore wind energy already contributes to electricity needs in Europe and Asia and now the first offshore wind farms are under development in U.S. waters. While the offshore wind industry has grown dramatically over the last several decades, especially in Europe, there is still a great deal of uncertainty associated with how the construction and operation of these wind farms affect marine animals and the marine environment.

Traditional offshore wind turbines are attached to the seabed in shallower waters within the nearshore marine environment. As offshore wind technologies become more advanced, floating structures have begun to be used in deeper waters where more wind resources exist.

Marine and Hydrokinetic Energy

Marine and Hydrokinetic (MHK) or marine energy development includes projects using the following devices:

- Wave power is the transport of energy by wind waves, and the capture of that energy to do useful work – for example, electricity generation or pumping water into reservoirs. A machine able to exploit significant waves in open coastal areas is generally known as a wave energy converter.
- Tidal power turbines are placed in coastal and estuarine areas and daily flows

are quite predictable.

- In-stream turbines in fast-moving rivers;
- Ocean current turbines in areas of strong marine currents;
- Ocean Thermal Energy Converters in deep tropical waters.

Thorium

Early thorium-based (MSR) nuclear reactor at Oak Ridge National Laboratory in the 1960s

Thorium is a fissionable material used in Thorium-based nuclear power. Proponents of thorium reactors claims several potential advantages over a uranium fuel cycle, such as thorium's greater abundance, better resistance to nuclear weapons proliferation, and reduced plutonium and actinide production. Thorium reactors can be modified to produce Uranium-233, which can then be processed into highly enriched uranium, which has been tested in low yield weapons, and is unproven on a commercial scale.

Investing in Alternative Energy

As an emerging economic sector, there are limited investment opportunities in alternative energy available to the general public. The public can buy shares of alternative energy companies from various stock markets, with wildly volatile returns. The recent IPO of SolarCity demonstrates the nascent nature of this sector- within a few weeks, it already had achieved the second highest market cap within the alternative energy sector.

Investors can also choose to invest in ETFs (exchange-traded funds) that track an alternative energy index, such as the WilderHill New Energy Index. Additionally, there are a number of mutual funds, such as Calvert's Global Alternative Energy Mutual Fund that are a bit more proactive in choosing the selected investments.

Recently, Mosaic Inc. launched an online platform allowing residents of California and New York to invest directly in solar. Investing in solar projects had previously been limited to accredited investors, or a small number of willing banks.

Over the last three years publicly traded alternative energy companies have been very volatile, with some 2007 returns in excess of 100%, some 2008 returns down 90% or more, and peak-to-trough returns in 2009 again over 100%. In general there are three subsegments of "alternative" energy investment: solar energy, wind energy and hybrid electric vehicles. Alternative energy sources which are renewable, free and have lower carbon emissions than what we have now are wind energy, solar energy, geothermal energy, and bio fuels. Each of these four segments involve very different technologies and investment concerns.

For example, photovoltaic solar energy is based on semiconductor processing and accordingly, benefits from steep cost reductions similar to those realized in the microprocessor industry (i.e., driven by larger scale, higher module efficiency, and improving processing technologies). PV solar energy is perhaps the only energy technology whose electricity generation cost could be reduced by half or more over the next 5 years. Better and more efficient manufacturing process and new technology such as advanced thin film solar cell is a good example of that helps to reduce industry cost.

The economics of solar PV electricity are highly dependent on silicon pricing and even companies whose technologies are based on other materials (e.g., First Solar) are impacted by the balance of supply and demand in the silicon market. In addition, because some companies sell completed solar cells on the open market (e.g., Q-Cells), this creates a low barrier to entry for companies that want to manufacture solar modules, which in turn can create an irrational pricing environment.

In contrast, because wind power has been harnessed for over 100 years, its underlying technology is relatively stable. Its economics are largely determined by siting (e.g., how hard the wind blows and the grid investment requirements) and the prices of steel (the largest component of a wind turbine) and select composites (used for the blades). Because current wind turbines are often in excess of 100 meters high, logistics and a global manufacturing platform are major sources of competitive advantage. These issues and others were explored in a research report by Sanford Bernstein. Some of its key conclusions are shown here.

Alternative Energy in Transportation

Due to steadily rising gas prices in 2008 with the US national average price per gallon of regular unleaded gas rising above $4.00 at one point, there has been a steady movement towards developing higher fuel efficiency and more alternative fuel vehicles for consumers. In response, many smaller companies have rapidly increased research and development into radically different ways of powering consumer vehicles. Hybrid and battery electric vehicles are commercially available and are gaining wider industry and consumer acceptance worldwide.

For example, Nissan USA introduced the world's first mass-production Electric Vehicle "Nissan Leaf". A plug-in hybrid car, the "Chevrolet Volt" also has been produced, using an electric motor to drive the wheels, and a small four-cylinder engine to generate additional electricity.

Making Alternative Energy Mainstream

Before alternative energy becomes mainstream there are a few crucial obstacles that it must overcome: First there must be increased understanding of how alternative energies work and why they are beneficial; secondly the availability components for these systems must increase; and lastly the pay-off time must be decreased.

For example, electric vehicles (EV) and Plug-in Hybrid Electric Vehicles (PHEV) are on the rise. These vehicles depend heavily on an effective charging infrastructure such as a smart grid infrastructure to be able to implement electricity as mainstream alternative energy for future transportations.

Research

There are numerous organizations within the academic, federal, and commercial sectors conducting large scale advanced research in the field of alternative energy. This research spans several areas of focus across the alternative energy spectrum. Most of the research is targeted at improving efficiency and increasing overall energy yields.

Multiple federally supported research organizations have focused on alternative energy in recent years. Two of the most prominent of these labs are Sandia National Laboratories and the National Renewable Energy Laboratory (NREL), both of which are funded by the United States Department of Energy and supported by various corporate partners. Sandia has a total budget of $2.4 billion while NREL has a budget of $375 million.

Mechanical Energy

Mechanical energy associated with human activities such as blood circulation, respiration, walking, typing and running etc. is ubiquitous but usually wasted. It has attracted tremendous attention from researchers around the globe to find methods to scavenge such mechanical energies. The best solution currently is to use piezoelecric materials, which can generate flow of electrons when deformed. Various devices using piezoelectric materials have been built to scavenge mechanical energy. Considering that the piezoelectric constant of the material plays a critical role in the overall performance of a piezoelectric device, one critical research direction to improve device efficiency is to find new material of large piezoelectric response. Lead Magnesium Niobate-Lead Titanate (PMN-PT) is a next-generation piezoelectric material with super high piezoelectric constant when ideal composition and orientation are obtained. In 2012, PMN-PT Nanowires with a very high piezoelectric constant were fabricated by a hydro-thermal approach and then assembled into an energy-harvesting device. The record-high piezoelectric constant was further improved by the fabrication of a single-crystal PMN-PT nanobelt, which was then used as the essential building block for a piezoelectric nanogenerator.

Solar

Solar energy can be used for heating, cooling or electrical power generation using the sun.

Solar heat has long been employed in passively and actively heated buildings, as well as district heating systems. Examples of the latter are the Drake Landing Solar Community is Alberta, Canada and numerous district systems in Denmark and Germany. In Europe, there are two programs for the application of solar heat: the Solar District Heating (SDH) and the International Energy Agency's Solar Heating and Cooling (SHC) program.

The obstacles preventing the large scale implementation of solar powered energy generation is the inefficiency of current solar technology, and the cost. Currently, photovoltaic (PV) panels only have the ability to convert around 16% of the sunlight that hits them into electricity.

Both Sandia National Laboratories and the National Renewable Energy Laboratory (NREL), have heavily funded solar research programs. The NREL solar program has a budget of around $75 million and develops research projects in the areas of photovoltaic (PV) technology, solar thermal energy, and solar radiation. The budget for Sandia's solar division is unknown, however it accounts for a significant percentage of the laboratory's $2.4 billion budget.

Several academic programs have focused on solar research in recent years. The Solar Energy Research Center (SERC) at University of North Carolina (UNC) has the sole purpose of developing cost effective solar technology. In 2008, researchers at Massachusetts Institute of Technology (MIT) developed a method to store solar energy by using it to produce hydrogen fuel from water. Such research is targeted at addressing the obstacle that solar development faces of storing energy for use during nighttime hours when the sun is not shining.

In February 2012, North Carolina-based Semprius Inc., a solar development company backed by German corporation Siemens, announced that they had developed the world's most efficient solar panel. The company claims that the prototype converts 33.9% of the sunlight that hits it to electricity, more than double the previous high-end conversion rate.

Wind

Wind energy research dates back several decades to the 1970s when NASA developed an analytical model to predict wind turbine power generation during high winds. Today, both Sandia National Laboratories and National Renewable Energy Laboratory have programs dedicated to wind research. Sandia's laboratory focuses on the advancement of materials, aerodynamics, and sensors. The NREL wind projects are centered on improving wind plant power production, reducing their capital costs, and making wind energy more cost effective overall.

The Field Laboratory for Optimized Wind Energy (FLOWE) at Caltech was established to research alternative approaches to wind energy farming technology practices that have the potential to reduce the cost, size, and environmental impact of wind energy production.

Renewable energies such as wind, solar, biomass and geothermal combined, supplied 1.3% of global final energy consumption in 2013.

Biomass

Sugarcane plantation to produce ethanol in Brazil

Biomass can be regarded as "biological material" derived from living, or recently living organisms. It most often refers to plants or plant-derived materials which are specifically called lignocellulosic biomass. As an energy source, biomass can either be used directly via combustion to produce heat, or indirectly after converting it to various forms of biofuel. Conversion of biomass to biofuel can be achieved by different methods which are broadly classified into: thermal, chemical, and biochemical methods. Wood remains the largest biomass energy source today; examples include forest residues (such as dead trees, branches and tree stumps), yard clippings, wood chips and even municipal solid waste. In the second sense, biomass includes plant or animal matter that can be converted into fibers or other industrial chemicals, including biofuels. Industrial biomass can be grown from numerous types of plants, including miscanthus, switchgrass, hemp, corn, poplar, willow, sorghum, sugarcane, bamboo, and a variety of tree species, ranging from eucalyptus to oil palm (palm oil).

A CHP power station using wood to supplies 30,000 households in France

Biomass, biogas and biofuels are burned to produce heat/power and in doing so harm the environment. Pollutants such as sulphurous oxides (SO_x), nitrous oxides (NO_x), and particulate matter (PM) are produced from this combustion. The World Health Organisation estimates that 7 million premature deaths are caused each year by air pollution, and biomass combustion is a major contributor of it. The use of biomas is carbon neutral over time, but is otherwise similar to burning fossil fuels.

Ethanol Biofuels

As the primary source of biofuels in North America, many organizations are conducting research in the area of ethanol production. On the Federal level, the USDA conducts a large amount of research regarding ethanol production in the United States. Much of this research is targeted toward the effect of ethanol production on domestic food markets.

The National Renewable Energy Laboratory has conducted various ethanol research projects, mainly in the area of cellulosic ethanol. Cellulosic ethanol has many benefits over traditional corn based-ethanol. It does not take away or directly conflict with the food supply because it is produced from wood, grasses, or non-edible parts of plants. Moreover, some studies have shown cellulosic ethanol to be more cost effective and economically sustainable than corn-based ethanol. Sandia National Laboratories conducts in-house cellulosic ethanol research and is also a member of the Joint BioEnergy Institute (JBEI), a research institute founded by the United States Department of Energy with the goal of developing cellulosic biofuels.

Other Biofuels

From 1978 to 1996, the National Renewable Energy Laboratory experimented with using algae as a biofuels source in the "Aquatic Species Program." A self-published article by Michael Briggs, at the University of New Hampshire Biofuels Group, offers estimates for the realistic replacement of all motor vehicle fuel with biofuels by utilizing algae that have a natural oil content greater than 50%, which Briggs suggests can be grown on algae ponds at wastewater treatment plants. This oil-rich algae can then be extracted from the system and processed into biofuels, with the dried remainder further reprocessed to create ethanol.

The production of algae to harvest oil for biofuels has not yet been undertaken on a commercial scale, but feasibility studies have been conducted to arrive at the above yield estimate. In addition to its projected high yield, algaculture— unlike food crop-based biofuels — does not entail a decrease in food production, since it requires neither farmland nor fresh water. Many companies are pursuing algae bio-reactors for various purposes, including scaling up biofuels production to commercial levels.

Several groups in various sectors are conducting research on Jatropha curcas, a poisonous shrub-like tree that produces seeds considered by many to be a viable source of biofuels feedstock oil. Much of this research focuses on improving the overall per acre oil yield of Jatropha through advancements in genetics, soil science, and horticultural practices. SG Biofuels, a San Diego-based Jatropha developer, has used molecular breeding and biotechnology to produce elite hybrid seeds of Jatropha that show significant yield improvements over first generation varieties. The Center for Sustainable Energy Farming (CfSEF) is a Los Angeles-based non-profit research organization dedicated to Jatropha research in the areas of plant science, agronomy, and horticulture. Successful exploration of these disciplines is projected to increase Jatropha farm production yields by 200-300% in the next ten years.

Geothermal

Geothermal energy is produced by tapping into the thermal energy created and stored within the earth. It is considered sustainable because that thermal energy is constantly replenished. However, the science of geothermal energy generation is still young and developing economic viability. Several entities, such as the National Renewable Energy Laboratory and Sandia National Laboratories are conducting research toward the goal of establishing a proven science around geothermal energy. The International Centre for Geothermal Research (IGC), a German geosciences research organization, is largely focused on geothermal energy development research.

Hydrogen

Over $1 billion has been spent on the research and development of hydrogen fuel in the United States. Both the National Renewable Energy Laboratory and Sandia National Laboratories have departments dedicated to hydrogen research.

Disadvantages

Further information: Disadvantages of hydroelectricity, Tidal power § Tidal power issues, and Issues relating to biofuels.

The generation of alternative energy on the scale needed to replace fossil energy, in an effort to reverse global climate change, is likely to have significant negative environmental impacts. For example, biomass energy generation would have to increase 7-fold to supply current primary energy demand, and up to 40-fold by 2100 given economic and energy growth projections. Humans already appropriate 30 to 40% of all photosynthetically fixed carbon worldwide, indicating that expansion of additional biomass harvesting is likely to stress ecosystems, in some cases precipitating collapse and extinction of animal species that have been deprived of vital food sources. The total amount of energy capture by vegetation in the United States each year is around 58 quads (61.5 EJ), about half of which is already harvested as agricultural crops and forest products. The remaining biomass is needed to maintain ecosystem functions and diversity. Since annual energy use in the United States is ca. 100 quads, biomass energy could supply only a very small fraction. To supply the current worldwide energy demand solely with biomass would require more than 10% of the Earth's land surface, which is comparable to the area use for all of world agriculture (i.e., ca. 1500 million hectares), indicating that further expansion of biomass energy generation will be difficult without precipitating an ethical conflict, given current world hunger statistics, over growing plants for biofuel versus food.

Given environmental concerns (e.g., fish migration, destruction of sensitive aquatic ecosystems, etc.) about building new dams to capture hydroelectric energy, further expansion of conventional hydropower in the United States is unlikely. Windpower, if deployed on the large scale necessary to substitute fossil energy, is likely to face public resistance. If 100% of U.S. energy demand were to be supplied by windmills, about 80 million hectares (i.e., more than 40% of all available farmland in the United States) would have to be covered with large windmills (50m hub height and 250 to 500 m apart). It is therefore not surprising that the major environmental impact of wind power is related to land use and less to wildlife (birds, bats, etc.) mortality. Unless only a relatively small fraction of electricity is generated by windmills in remote locations, it is unlikely that the public will tolerate large windfarms given concerns about blade noise and aesthetics.

Biofuels are different from fossil fuels in regard to net greenhouse gases but are similar to fossil fuels in that biofuels contribute to air pollution. Burning produces airborne carbon particulates, carbon monoxide and nitrous oxides.

References

- Norman F. Cantor (1993). The Civilization of the Middle Ages: The Life and death of a Civilization. Harper Collins. p. 564. ISBN 978-0-06-092553-6.
- Krotz, Dan (2012-05-13). "Berkeley Lab Scientists Generate Electricity From Viruses | Berkeley Lab". News Center. Retrieved 2016-08-14.
- "Pathways to a Low-Carbon Economy: Version 2 of the Global Greenhouse Gas Abatement Cost Curve". McKinsey Global Institute: 7. 2009. Retrieved February 16, 2016.
- The Energy of the Future: Fourth "Energy Transition" Monitoring Report — Summary (PDF).

Berlin, Germany: Federal Ministry for Economic Affairs and Energy (BMWi). November 2015. Retrieved 2016-06-09.

- "National Action Plan on Energy Efficiency (NAPE): making more out of energy". Federal Ministry for Economic Affairs and Energy (BMWi). Retrieved 2016-06-07.
- Making more out of energy: National Action Plan on Energy Efficiency (PDF). Berlin, Germany: Federal Ministry for Economic Affairs and Energy (BMWi). December 2014. Retrieved 2016-06-07.
- "Gabriel: Efficiency First — discuss the Green Paper on Energy Efficiency with us!" (Press release). Berlin, Germany: Federal Ministry for Economic Affairs and Energy (BMWi). 12 August 2016. Retrieved 2016-09-06.
- Amelang, Sören (15 August 2016). "Lagging efficiency to get top priority in Germany's Energiewende". Clean Energy Wire (CLEW). Berlin, Germany. Retrieved 2016-09-06.
- Huntington, Hillard (2011). EMF 25: Energy efficiency and climate change mitigation — Executive summary report (volume 1) (PDF). Stanford, CA, USA: Energy Modeling Forum. Retrieved 2016-05-10
- Environmental and Energy Study Institute. "Industrial Energy Efficiency: Using new technologies to reduce energy use in industry and manufacturing" (PDF). Retrieved 2015-01-11.
- "The City's Evolution–From Old Dartmouth to New Bedford, Whaling Metropolis of the World". Old Dartmouth Historical Society. Retrieved 2015-01-30.
- McKenna, Paul (5 December 2012). "Is the "Superfuel" Thorium Riskier Than We Thought?". Popular Mechanics. Retrieved 2 March 2015.
- T.A. Volk; L.P. Abrahamson (January 2000). "Developing a Willow Biomass Crop Enterprise for Bioenergy and Bioproducts in the United States". North East Regional Biomass Program. Retrieved 4 June 2015.
- Matar, W (2015). "Beyond the end-consumer: how would improvements in residential energy efficiency affect the power sector in Saudi Arabia?". Energy Efficiency. doi:10.1007/s12053-015-9392-9.
- Okulski, Travis (June 26, 2012). "Audi's Carbon Neutral E-Gas Is Real And They're Actually Making It". Jalopnik (Gawker Media). Retrieved 29 July 2013.
- Elisabeth Rosenthal (2007-01-31). "Once a Dream Fuel, Palm Oil May Be an Eco-Nightmare". New York Times. Archived from the original on February 11, 2012. Retrieved 2008-12-14.

Permissions

All chapters in this book are published with permission under the Creative Commons Attribution Share Alike License or equivalent. Every chapter published in this book has been scrutinized by our experts. Their significance has been extensively debated. The topics covered herein carry significant information for a comprehensive understanding. They may even be implemented as practical applications or may be referred to as a beginning point for further studies.

We would like to thank the editorial team for lending their expertise to make the book truly unique. They have played a crucial role in the development of this book. Without their invaluable contributions this book wouldn't have been possible. They have made vital efforts to compile up to date information on the varied aspects of this subject to make this book a valuable addition to the collection of many professionals and students.

This book was conceptualized with the vision of imparting up-to-date and integrated information in this field. To ensure the same, a matchless editorial board was set up. Every individual on the board went through rigorous rounds of assessment to prove their worth. After which they invested a large part of their time researching and compiling the most relevant data for our readers.

The editorial board has been involved in producing this book since its inception. They have spent rigorous hours researching and exploring the diverse topics which have resulted in the successful publishing of this book. They have passed on their knowledge of decades through this book. To expedite this challenging task, the publisher supported the team at every step. A small team of assistant editors was also appointed to further simplify the editing procedure and attain best results for the readers.

Apart from the editorial board, the designing team has also invested a significant amount of their time in understanding the subject and creating the most relevant covers. They scrutinized every image to scout for the most suitable representation of the subject and create an appropriate cover for the book.

The publishing team has been an ardent support to the editorial, designing and production team. Their endless efforts to recruit the best for this project, has resulted in the accomplishment of this book. They are a veteran in the field of academics and their pool of knowledge is as vast as their experience in printing. Their expertise and guidance has proved useful at every step. Their uncompromising quality standards have made this book an exceptional effort. Their encouragement from time to time has been an inspiration for everyone.

The publisher and the editorial board hope that this book will prove to be a valuable piece of knowledge for students, practitioners and scholars across the globe.

Index

A

Ac/ac Converters, 41, 43-44

Alternative Energy, 9, 221, 245-249, 253-255, 259

B

Balanced Systems, 220

Biological Hydrogen Production, 251

Biomass Briquettes, 250

Bulk Power Transmission, 168

C

Capacitive, 34, 49, 52

Coal As An Alternative To Wood, 246

Coal Gasification As An Alternative To Petroleum, 247

Copper Oxide Rectifiers, 72

Crystal Detector, 54, 71

D

Dc Power Source Usage, 79

Dc-to-dc Converter, 30, 47, 50, 54

Dc/ac Converters (inverters), 33, 44

Demolition, 111

E

Efficient Energy Use, 221, 229, 241-242

Electric Motor Speed Control, 79

Electric Power Distribution, 11, 79, 143, 153-155, 161, 211, 214

Electric Power System, 143, 145, 147, 149, 151, 153, 155, 157, 159, 161, 163, 165, 167, 169, 171, 173, 175, 177, 179, 181, 183, 185, 187, 189, 191, 193, 195, 197, 199, 201, 203, 205, 207, 209, 211, 213

Electric Power Transmission, 143, 161, 166-167, 172, 179, 202

Electrical Engineering, 1-3, 11-14, 16, 19, 22-24, 26-29, 144, 214, 216

Electrical Generator, 102-103, 107, 118

Electrical Transformer, 119

Electrochemistry, 121, 224

Electrolytic, 68

Electromagnet Laminations, 120

Electromagnetic Induction, 2, 13, 90, 113-115, 117, 188, 207, 222-223

Electromechanical Conversion, 52

Electronic Components, 18-19, 41, 89, 93

Electronic Conversion, 49

Electronics, 5, 11-12, 14, 16-17, 19, 22, 24, 26-27, 29-33, 35, 37, 39, 41-47, 49, 51, 53, 55, 57, 59, 61, 63, 65, 67, 69, 71, 73, 75-77, 79, 81, 83, 85, 87, 89, 92-93, 95, 107, 120, 146, 149-150, 214

Electroshock Weapons, 81

Energy Engineering, 1-28, 30, 32, 34, 36, 38, 40, 42, 44, 46, 48, 50, 52, 54, 56, 58, 60, 62, 64, 66, 68, 70, 72, 74, 76, 78, 80, 82, 84, 86, 88, 92, 94, 96, 98, 100, 102, 104, 106, 108, 110, 112, 114, 116, 118, 120, 122, 124, 126, 128, 130, 132, 134, 136, 138, 140, 142, 144, 146, 148, 150, 152, 154, 156, 158, 160, 162, 164, 166, 168, 170, 172, 174, 176, 178, 180, 182, 184, 186, 188, 190, 192, 194, 196, 198, 200, 202, 204, 206, 208, 210, 212, 214, 216, 218, 220-261

Energy Losses, 48, 135, 168, 189, 195, 229, 238

Environmental Impacts of Photovoltaic Technologies, 129

Ethanol as an Alternative to Fossil Fuels, 246

Ethanol Biofuels, 258

F

Fault Current Limiter, 143, 211-213

Fueling Operations, 95

Full-wave Rectification, 56-57, 64

G

Generation, 1, 4-8, 14, 16, 25, 47, 81, 90, 95, 99, 124-125, 128-130, 133-134, 141, 145, 155-156, 163, 165-166, 168, 177-179, 181, 183, 207, 212, 214, 216, 221-229, 234, 238, 247, 251-252, 254-256, 259-260

Generator Sets, 52, 86, 185

Geothermal, 9, 169, 221-223, 227, 229, 245, 247-248, 253, 257, 259

H

Half-wave Rectification, 55-56, 64

High-voltage Transmission Network, 185

Horizontal Axis, 98, 102-103, 105-106, 125

Hvdc Power Transmission, 81, 85, 87

Hydroelectricity, 226, 251, 259

Hydrokinetic Energy, 252

I
Ideal Transformer, 174, 189-194
Induction Coils, 208-209, 211
Induction Heating, 80
Inductive Fault Current Limiter, 213
Input Voltage, 8, 35, 41, 49-50, 52-54, 56, 59, 62-64, 74, 88
Insulation Drying, 204

L
Low-voltage Networks, 186

M
Magnetic Flow Meter, 120
Mechanical Energy, 9, 118-119, 223, 225, 255
Merchant Transmission, 180
Motor-generator Set, 52, 67-68
Multilevel Inverters, 33, 40-41, 85

N
Nuclear Transformation, 225

O
One-line Diagram, 215-216, 219-220
Operability, 111
Output Frequency, 42, 77
Output Power, 30, 42, 52, 63, 78, 126
Output Voltage, 35-37, 42, 44, 48, 50, 52-56, 58-60, 62, 64-65, 71, 74-75, 78-81, 84, 88, 126, 205, 208
Output Waveform, 33-37, 75, 80, 85-86
Ozone Cracking, 95-96

P
Parasitic Induction Within Conductors, 121
Petroleum as an Alternative To Whale Oil, 246
Photovoltaic Effect, 121, 124, 224
Photovoltaics, 45, 121, 123-128, 132-133, 135-138, 140-142, 169, 221, 225
Piezoelectric Effect, 225
Power Electronics, 5, 12, 16, 19, 27, 29-31, 33, 35, 37, 39, 41-47, 49, 51, 53, 55, 57, 59, 61, 63, 65, 67, 69, 71, 73, 75, 77, 79, 81, 83, 85, 87, 89, 146, 149-150, 214
Power Engineering, 1-6, 10, 12, 16, 144, 183, 215, 217, 219-220
Power Inverter, 29, 74, 76-78, 86, 88
Power-flow Problem Formulation, 216
Power-flow Study, 215-216
Power-system Protection, 143, 183
Prevention of Static Electricity, 92
Primary Distribution, 153-154, 157, 159

R
Real Transformer, 191-195
Rectifier, 4, 29-31, 38, 43-45, 47, 49, 54-73, 79, 87-88, 144, 150, 205
Rectifier Circuits, 55, 57, 61, 87-88
Rectifier Devices, 55
Rectifier Output Smoothing, 64
Repowering, 111

S
Secondary Distribution, 154, 159
Signal Processing, 12, 19
Single Wire Earth Return, 158, 182
Single-phase Full-bridge Inverter, 34, 36
Single-phase Half-bridge Inverter, 34
Single-phase Rectifiers, 55, 57, 61
Small Wind Turbines, 101, 110
Solar Cells, 73, 122, 124-127, 130-131, 133-137, 140-141, 224, 254
Solid State Fault Current Limiter, 213
Solid-state Transistors, 16
Static Discharge, 92-95
Static Electricity, 13, 90-98, 222
Subtransmission, 154, 164, 172-173
Superconducting Cables, 182
Superconducting Fault Current Limiter, 212
Synchronous Rectifier, 49, 67

T
Telecommunications, 12, 19-20, 26, 28, 179
Thermoelectric Effect, 225
Three-phase Bridge Rectifier, 59
Three-phase Rectifiers, 57, 63, 88
Three-phase Voltage Source Inverter, 34, 37
Three-phase, Half-wave Circuit, 58
Transformer, 3, 8, 14, 43, 47-52, 56-58, 61-67, 71, 75, 77, 81-83, 85, 94, 114, 117, 119, 143-144, 152, 154-159, 161, 166, 169-170, 173-174, 184-185, 188-211, 213-215, 219-220
Transmission, 1-3, 5-8, 15-16, 20, 26-27, 30-31, 33, 44, 47, 51, 55, 59, 68-69, 73, 75, 81, 85, 87-89,

104, 107-108, 141, 143-145, 147-149, 153, 155-157, 161-186, 189, 197, 199, 202, 205-207, 209, 211, 213-214, 216-217, 221-222, 228
Turbine Monitoring, 108
Twelve-pulse Bridge, 61

U

Unbalanced Systems, 220
Unconventional Designs, 108
Underground Transmission, 148, 163, 165
Uninterruptible Power Supplies, 79, 85
Utilization, 1, 5-6, 8-9, 153, 189, 205, 210, 232

V

Vertical Axis, 99, 102, 104-106, 110-112
Vibrating Rectifier, 67
Voltage-multiplying Rectifiers, 62

W

Wind Turbine, 44, 98-103, 105-113, 150, 224, 254, 256
Wind Turbines, 46, 48, 90, 98-99, 101-102, 104-113, 146, 150, 178, 226, 245, 252, 254
Wireless Power Transmission, 182-183